T0361802

From Molecular Genetics to Genomics

With the rise of genomics, the life sciences have entered a new era. Maps of genomes have become the icons for a comprehensive knowledge of the organism on a previously unattained level of complexity. This book provides an in-depth history of molecular genetics and genomics.

Part I of the book shows how the cartography of classical genetics was linked to the molecular analysis of gene structure through the introduction of new model organisms, such as bacteria, and through the invention of new experimental tools, such as gene transfer. Part II addresses the development of human genome sequencing in all its technical, epistemic, social, and economic complexity.

With its detailed analyses of the scientific practices and its illustration of the diversity of mapping, this book is a significant contribution to the history of genetics.

A companion volume from the same editors—*Classical Genetic Research and its Legacy: The Mapping Cultures of Twentieth-century Genetics*—covers the history of mapping procedures as they were developed in classical genetics.

Jean-Paul Gaudillière is a senior researcher at the National Institute of Health (INSERM), Paris. His work has addressed many aspects of the history of molecular biology and of the biomedical sciences during the twentieth century. His actual research focuses on the history of biological drugs. He is the co-editor of *Heredity and Infection: A History of Disease Transmission* (2002).

Hans-Jörg Rheinberger is Director at the Max Planck Institute for the History of Science in Berlin. He has published numerous papers and books on molecular biology and the history of science, including a co-edited collection, *The Concept of the Gene in Development and Evolution* (2000).

Routledge studies in the history of science, technology and medicine

Edited by John Krige

Georgia Institute of Technology, Atlanta, USA

Routledge Studies in the History of Science, Technology and Medicine aims to stimulate research in the field, concentrating on the twentieth century. It seeks to contribute to our understanding of science, technology and medicine as they are embedded in society, exploring the links between the subjects on the one hand and the cultural, economic, political and institutional contexts of their genesis and development on the other. Within this framework, and while not favouring any particular methodological approach, the series welcomes studies which examine relations between science, technology, medicine and society in new ways, e.g. the social construction of technologies, large technical systems etc.

From Molecular Genetics to Genomics

The mapping cultures of twentieth-century genetics

Edited by
Jean-Paul Gaudillière and
Hans-Jörg Rheinberger

Routledge
Taylor & Francis Group

LONDON AND NEW YORK

First published 2004
by Routledge
2 Park Square, Milton Park, Abingdon, Oxon OX14 4RN

Simultaneously published in the USA and Canada
by Routledge
270 Madison Ave, New York, NY 10016

Routledge is an imprint of the Taylor & Francis Group

Transferred to Digital Printing 2005

Typeset in Baskerville MT by
Newgen Imaging Systems (P) Ltd, Chennai, India

British Library Cataloguing in Publication Data
A catalogue record for this book is available
from the British Library

Library of Congress Cataloging in Publication Data
A catalog record for this book has been requested

ISBN 0–415–32850–0

Printed and bound by Antony Rowe Ltd, Eastbourne

This book is dedicated to the memory of
Frederic L. Holmes

Contents

Commentaries   201

Figures

Notes on contributors

Adam Bostanci is a Doctoral Research Student at the ESRC Centre for Genomics in Society at the University of Exeter, where his research focuses on the laboratory practices and international politics of genome sequencing. Previously, he has worked as a science journalist, contributing to daily papers, popular science magazines and the research weekly *Science*.

Soraya de Chadarevian is Senior Research Associate and Affiliated Lecturer at the Department of History and Philosophy of Science at the University of Cambridge. Recent publications include *Designs for Life. Molecular Biology after World War II* (2002) and two co-edited volumes, *Molecularizing Biology and Medicine: New Practices and Alliances, 1910s–1970s* (1998) and *Models: The Third Dimension of Science* (2004)

Angela N.H. Creager is Associate Professor in the Department of History and Program in History of Science at Princeton University. She is author of *The Life of a Virus: Tobacco Mosaic Virus as an Experimental Model, 1930–1965* (2002) and co-editor of two volumes, *Feminism in Twentieth-Century Science, Technology and Medicine* (2001), with E. Lunbeck and L. Schiebinger, and *The Animal–Human Boundary: Historical Perspectives* (2003), with W. Chester Jordan.

Jean-Paul Gaudillière is a senior researcher at INSERM, Paris. His work has addressed many aspects of the history of molecular biology and of the bio-medical sciences during the twentieth century. His actual research focuses on the history of biological drugs. He is the co-editor of *Heredity and Infection: A History of Disease Transmission* (2002) and the author of *Inventer la biomédecine. La France, l'Amérique et la production des savoirs du vivant (1945–1965)* (2002).

Scott F. Gilbert is the Howard A. Schneidermann Professor of Biology at Swarthmore College. He is the author of *Developmental Biology* (now in its seventh edition) and the editor of *A Conceptual History of Modern Embryology* (1991). His research concerns the developmental origins of evolutionary novelties, and the relationships between embryology, evolution, and ecology.

David Gugerli holds the Chair for the History of Technology at the Swiss Federal Institute of Technology (ETH) in Zurich. One of his books is concerned with the discursive currents of electrification (*Redeströme*, 1996), the most recent study deals with mapping, politics, and landscape in the nineteenth century (*Topografien der Nation*, 2002, co-authored with D. Speich).

Stephen Hilgartner is Associate Professor in the Department of Science & Technology Studies at Cornell University. His recent book, *Science on Stage: Expert Advice as Public Drama* (2000), won the Carson Prize from the Society for Social Studies of Science.

Frederic L. Holmes was Professor in the Department of History of Science and Medicine at Yale University. His publications span the history of biology, biochemistry, physiology, and chemistry. His particular interest was in the fine structure of creative scientific activity. Among his books are *Claude Bernard and Animal Chemistry* (1974), *Lavoisier and the Chemistry of Life* (1985), *Hans Krebs* (1991/93) and *Meselson, Stahl, and the Replication of DNA* (2001).

Alain Kaufmann is Director of the "Science, Medicine and Society Interface" of the University of Lausanne where he is in charge of the interface between the research done in his university and science museums, schools and the general public. His publications include "Transferts de modèles entre les sciences de la vie et l'informatique: le cas des algorithmes génétiques. Une approche sociologique", *Computer Science, Communications and Society: A Technical and Cultural Challenge* (with Francesco Panese, Lausanne 1993); "The Affair of the Memory of Water: Towards a Sociology of Scientific Communication" (*Réseaux—The French Journal of Communication*, 1994); "Recherche, enseignement et culture scientifique: pistes de réflexion pour une régulation réciproque", Swiss National Commission for the UNESCO and Swiss Academy for Natural Sciences, 2001.

Gísli Pálsson is Professor of Anthropology at the University of Iceland and the University of Oslo. His main books are *Writing on Ice: The Ethnographic Notebooks of Vilhjalmur Stefansson* (2001, editor and Introduction), *The Textual Life of Savants: Ethnography, Iceland and the Linguistic Turn* (1995), *Nature and Society: Anthropological Perspectives* (1996, co-edited with Philip Descola), and *Images of Contemporary Iceland* (1996, co-edited with E.P. Durrenberger).

Hans-Jörg Rheinberger is Director at the Max Planck Institute for the History of Science in Berlin. He has published numerous articles on molecular biology and the history of science. He has written, among other books, *Toward a History of Epistemic Things. Synthesizing Proteins in the Test Tube* (1997); ed. (together with P. Beurton and R. Falk), *The Concept of the Gene in Development and Evolution* (2000); *Experimentalsysteme und epistemische Dinge* (2001); ed. (together with F.L. Holmes and J. Renn), *Reworking the Bench. Research Notebooks in the History of Science* (2003).

Sergio Sismondo teaches philosophy and sociology of science at Queen's University, Canada. He is the author of *An Introduction to Science and Technology Studies* (2004) and co-author with B. Castel of *The Art of Science* (2003).

Marcel Weber is Swiss National Science Foundation Professor of Philosophy of Science at the University of Basel, Switzerland. Among his books are *Philosophy of Experimental Biology* (in press) and *Die Architektur der Synthese. Entstehung und Philosophie der modernen Evolutionstheorie* (1998).

Acknowledgments

We wish to thank all the contributors to this volume for sharing their expertise and having been so cooperative and patient in the process of finalizing the manuscript. We would also like to express our gratitude to Gabriele Brüß and Antje Radeck for helping with the difficult task of making everything ready for the press. We thank John Krige as the Series editor for his interest, and two anonymous referees for their valuable comments. Finally, we would like to acknowledge Jo Whiting, Terry Clague, and Amritpal Bangard from Routledge and Vincent Antony from the Production Office for guiding us so smoothly through the production stages.

The editors
Jean-Paul Gaudillière
Hans-Jörg Rheinberger

Introduction

Jean-Paul Gaudillière and Hans-Jörg Rheinberger

This book is about the mapping cultures of twentieth-century genetics. *From Molecular Genetics to Genomics* is the second volume of a collection of papers resulting from a conference that was held at the Max Planck Institute for the History of Science in Berlin in March 2001. It covers the mapping procedures developed in the era of molecular biology, gene technology, and genomics, that is, roughly, in the second part of the twentieth century. The accompanying first volume—*Classical Genetic Research and its Legacy*—covers the first part of the twentieth century. We felt that a detailed history of genetic mapping was timely, if not overdue, in view of the recent developments in genomics epitomized by the announcement of a first draft of the structure of the human genome in the summer of 2000, which was eventually made public in February 2001, a month before the workshop from whose deliberations this book took shape.

It is possibly not too far-fetched to postulate that the spaces of our genetic knowledge about living beings have been organized around two major fields of metaphor, metonymy, and model with their concomitant practices. One of these fields gravitates around the theme of mapping, the other around that of information and communication. Whereas the concept of information has been given wide attention by historians and philosophers of science, particularly as far as the history of molecular biology is concerned, mapping has not. This book is about mapping. Both themes have not only transformed biology, they have also created powerful public images and icons associated with contemporary genomics, its potential uses and abuses, that is, the genetic map and the genetic program. Whereas the notion of information has become firmly embedded within the space once defined and confined by quite different conceptions of biological specificity and function, the idea of the map resides and exerts its power predominantly at the level of structure, organization, and correlation. And whereas mapping is clearly associated with the activity of organizing and of performing a certain kind and form of research, with its instruments and practices circulating in specific communities—that is, with a distinctly epistemological and performative meaning—the concept of information carries the burden of the new molecular genetic ontology. It should be an interesting exercise to ask why these two metaphorical realms could obviously coexist in such a productive manner and why they could develop such an enormous impact on the representation and the manipulation of life over the past century, although they do not have any obvious theoretical linkage to each other.

The year 1953 was remarkable in the history of the life sciences. It was the year that Francis Crick and James Watson first presented their model of the double helical structure of DNA, to whose fiftieth anniversary this book is a contribution. With it, the way was open to define the classical gene on the molecular level. It was a turning point for the molecular genetic meaning of information, and for genetic mapping as well. Historians have discussed many dimensions of the post-war molecularization of the life sciences. Revisiting it from the perspective of mapping practices—as this volume does—proves valuable as it helps to prob-lematize the conventional conception of a direct line supposedly linking chromo-somal genetics, postwar bacterial genetics, and the discovery of DNA. Mapping actually played a critical role in the molecularization of the gene as testified here, but the techniques used were not a direct continuation of classical genetic map-ping. Further, molecularizing the gene eventually led to the recent mapping of the human genome; but again the connections are far from direct. Classical mapping was superseded in the sense that it became a problem of physical and chemical analysis. Mapping was also complemented by and juxtaposed with sequencing. As shown in this volume, these new practices did not emerge out of a continuation of classical genetics, but from another horizon, from the world of animal models and human genetics, typical of postwar biomedicine.

The year 1953 was also a remarkable year for mapping in the philosophy of science, albeit perhaps less spectacularly, and certainly less well known than the advent of the double helix. In 1953, a British philosopher in the tradition of Ludwig Wittgenstein, Stephen Toulmin, published a little book with a very ambi-tious title: *The Philosophy of Science: An Introduction*. In it, Toulmin tells us that in physics as elsewhere in science, we draw consequences by drawing lines (Toulmin 1953: 25). In a chapter entitled "Theories and Maps," Toulmin extended this gen-eral epistemological claim. He argued that the comparison of scientific laws with generalizations, and the logical structure in which generalizations are framed, had for a long time misled philosophers of science in their search for models of scientific activity. Instead, he postulated, physical theories should be compared with maps. He contended that diagrams and equations could be seen as a means for setting up "maps of phenomena" (Toulmin 1953: 108), and that the mapping metaphor could be used to shed fresh light on dark and dusty corners and impasses in the philosophy of science.

Maps, for those who know to read them are, first of all, a means of orientation. But they also condense and depict phenomena in relation to each other. Thus, they create meaning through condensation and relating—that is, through the power of synopsis—as much as through representation and reference—that is, through the ongoing traffic between the map and the mapped territory. It has become a common observation in the history of science that models and tools have a life of their own and may profoundly transform the phenomena they are supposed to represent. As Georges Canguilhem once wrote about "life" in the laboratory: "We must not forget that the laboratory itself constitutes a new environment in which life certainly establishes norms whose extrapolation does not work without risk when removed from the conditions to which these norms relate. For the animal and for man the laboratory environment is one possible environment among others (...) for the living being apparatuses and products are the objects among which it moves as in an unusual world. It is impossible that the ways of life in the laboratory should fail to retain any specificity in their relationship to the place and moment of the experiment" (Canguilhem 1966, quoted from 1991: 148).

The transformational role of models and tools is especially true for maps in general and genetic maps in particular. As discussed in the previous volume of this history of mapping this form of biological research could only take off when geneticists worked out a strong linkage between the vision of hereditary factors as being linearly arranged on chromosomal threads, and the breeding procedures for producing standardized strains of organisms. Since their arrival genetic maps have delivered powerful images of the relations between things, within and outside the confines of the laboratory. Genetic items have been related, not only to each other, but also to matters technical, economical, or political. Yet this diversity was kept under control. The heterogeneity of purposes, organisms, and epistemic entities was homogenized to varying degrees, thus giving birth to the mapping culture that constitutes the common background to the processes described and analyzed in this book. The basic mapping conception of linearly arranged genetic elements proved more robust than any of its early advocates might have thought. But it also proved deceptive, channeling huge research energies into a one-dimensional genetic vision which supposedly accounted for the entire organism.

Since the days of the Drosophila group, mapping procedures have proliferated extensively. One might even be tempted to associate—if not to match—the punctuated history of gene concepts throughout the latter part of the last century with the history of particular mapping and sequencing practices, which themselves resonated with and were instrumental in creating broader cultural environments. Following this diversity through the development of molecular genetics and genomics, this volume intends to address, throughout its different sections, four sets of more general, closely related issues.

The first set of questions concerns the *epistemic dimension* of genetic mapping as a way of "spatializing" knowledge about transmission, inheritance, and biological traits. Mapping procedures are devices that reduce dimensionality and thereby increase orientation, creating what Bruno Latour called "immutable mobiles" able to carry on accepted genetic knowledge over space and time (Latour 1990). The reducing power of maps also inevitably creates fictions of homogeneity by squeezing all available information into one representational space. Mapping procedures in general have a long history in Western science and culture as ways of representing, orienting, or navigating, in geography as well as in taxonomy. Whereas geographical maps are usually two-dimensional representations, taxonomic orders are typically used to "map" the diversity of organisms into ladders (such as the Great Chain of Being), landscapes (such as Linné's vision of a natural system), or tree-like structures (such as Darwin's topology of descendence). Genetic maps tend to be one-dimensional arrays. This tendency was strongly reinforced as chromosomal analysis identified fragments of DNA strings used as genetic markers, or sequences of nucleotides obtained from nucleic acids. Given this one-dimensionality, one might wonder whether such genetic arrays can properly be viewed as maps at all. The reduction to one dimension also raises questions about the historical relationship between genetics and statistics, and the emergence of alternative procedures in the history of genetics and contemporary biology. Other ways of arranging genes based on different procedures for breeding and crossing were actually proposed, albeit rarely. One aspect of an encompassing history of genetic mapping is to understand how and why these soon left the scene. One may view the general impact of this one-dimensionality in the larger context of the operational reductionism of the modern sciences, in particular molecular biology. One final epistemological conundrum is tied to the perennial question of the relations between representation and intervention, or more specifically, the relation between maps as displaying genetic patterns and maps as tools for creating such patterns, and ultimately paving the way to new phenomena. *Molecularizing Maps*, the first part of the present book, deals with the transition from classical to molecular genetics, and offers rich material for studying the connections between constructive and representational aspects of science in the making.

The second set of questions focuses on mapping as a form of *biological work*. Genetic mapping was from its very beginning a highly integrative and technologically oriented activity. Construction patterns however changed with the organismal, instrumental or cultural contingencies that shaped the communities of geneticists. Investigating mapping as a system of actions involves looking at the various ways in which these communities have handled and linked the procurement of resources, the assignment of experimental tasks, and the dissemination of results. The selection and construction of biological tools, the establishment and organization of reference centers, the various models and representational devices used, the forms of exchange and appropriation within different mapping collectives, and the social and political environments of mapping centers, fall within different conjunctions. The case studies of this volume, especially in the part on the *Moral and Political Economy of Human Genome Sequencing*, give a fair impression of the fact that we must take into account the widely varying arrangements of these factors over time. Mapping work has accordingly been achieved in settings like research laboratories, resource centers, industrial research facilities and, more recently, biotechnology startups. The emergence of the latter has attracted attention to a last dimension of mapping systems, namely the development of exchange economies amending the gift system of the "lords of the fly" to the commercial circulation of appropriated—eventually patented—biological material.

A third question concerns the *changing scale* and the changing social nature of genetic research during the twentieth century. As documented—especially in the chapters of the second part—scaling up is reflected in the increasing complexities of mapping networks. These complexities are illustrated by changing data exchange regimes, from the *Drosophila Newsletter*, through the form of collective experimentation, which characterized the mapping of the human histocompatibility genetic complex in the 1970s, to the recent genomic consortia. Complexity is however just one aspect of big mapping. Other aspects analyzed in this volume are the growing reliance on technology, and the impact mapping made on the molecularization of life. The increasing technological component of mapping is easy to perceive when placing side by side the milk bottles and microscopes of the Drosophila geneticists, the Petri dishes and ultracentrifuges employed by phage and bacterial geneticists, and the sequencing platforms of the worm and the human genome people. The changing nature and roles of instrumentation thus raise issues about the emergence of standardized mapping tools: about their local or global circulation, about their differential uses, and about the consequences of their production.

Fourth, comes the issue of *diversity*. It is a question of historical inquiry taken up in the contributions to this volume, and the accompanying volume on *Classical Genetic Research and its Legacy*, to decide whether it is justified to write about one single mapping culture to characterize genetics in the twentieth century. The chapters in this book highlight the similarities and discrepancies between the mapping practices that developed in various areas of biological research and at different times. The two volumes give an overview of this diversity, including analyses of different model organisms (bacteria, phage, humans), different specialities (human genetics, immunology, biochemistry), and different agendas (cytological mapping, mapping human Mendelian pathological traits, mapping biochemical pathways, mapping chromosomes by means of bacterial conjugation, and finally sequencing entire genomes). Not least, attention must be given to the different forms in which maps can come and circulate, including lists, inventories, plans, diagrams, pictures, and databases. It thus appears that the various components of the *Drosophila* exemplar were by no means linearly and directly passed on, or merely echoed in the various domains where mapping came to be used as a means of relating pertinent phenomena to each other.

This volume is organized in two parts. The first concentrates on the molecularization of the gene. It gathers studies of the various ways in which microorganisms like phages (Frederic L. Holmes) and bacteria (Angela N.H. Creager) were assigned genes and chromosomes, which eventually became targets of mapping enterprises. The molecularization of the gene was however not limited to linking mapping techniques with the model organisms selected by the postwar molecular biologists. The process changed the nature of classical Drosophila genetics and other domains of the biological sciences. This proved especially important for embryology, which had a strong tradition of spatial representation. New applications thus connected genetic maps with biochemical charts (Angela N.H. Creager), developmental maps (Marcel Weber, Scott F. Gilbert), cellular fate maps, or neurological maps (Soraya de Chadarevian).

The second part of the volume focuses on late-twentieth-century genome research. This development affected the world of genetics in so many ways that one may argue that a sequencing culture has taken over the mapping cultures.

Sequences are however far from being the only products of genome research. Rather than becoming obsolete, maps are more numerous than ever, and their producers and users are no longer just geneticists. Mapping in the genome era is rooted in moral and political systems, which differ significantly from those of the "lords of the fly." Focusing on the production and uses of sequences, this volume approaches these systems in various ways. Laboratory practices, and the relationship between maps and sequences, are discussed in Adam Bostanci's chapter. New actors and new ways of organizing research are at the center of Stephen Hilgartner and Alain Kaufmann's chapters, which both follow the activities of integrated genome research centers. The appropriation of genetic collections and genetic data has become a controversial issue in many mapping and sequencing projects including the Icelandic Decode initiative. Echoing and enforcing the main theme of this book, Gísli Pálsson analyzes the problematic relations of this enterprise with older visions of human and medical genetics. These diverse themes could be extended. They stand as an indication of a historical transformation that will never be fully understood if only seen from the perspective of one of its isolated realizations. The two commentaries at the end of the book by philosopher Sergio Sismondo and historian of technology David Gugerli reinforce this point on a more general level.

From this perspective, one final remark should be made. Taking biotechnology as an example, observers of contemporary science often argue that a new "mode of knowledge production" has emerged (Gibbons *et al.* 1994). The new world of science is seen as project-oriented, multidisciplinary, supported by heterogeneous partners, combining basic and applied research, as well as being more closely related to the markets. Taking into account the *longue durée* of biological research suggests that all sorts of continuities underpin this revolution. It is possibly the greatest challenge for a book like this to strike a balance between identifying permanent and transient aspects. It is the historian's usual task to read the past within the past. The preparation of this book has however convinced us that a retrospective gaze is not always detrimental to historical writing. Without the benefit of a genomic-informed perspective, one might doubt that "mapping" would have become a topic considered to be critical to the history of twentieth-century biology.

Bibliography

Canguilhem, G. (1966) *Le normal et le pathologique*, Paris: Presses Universitaires de France; English translation (1991) *The Normal and the Pathological*, New York: Zone Books.

Gibbons, M., Limoges, C., Nowotny, H., Schwartzmann, S., Scott, P. and Trow, M. (1994) *The New Production of Knowledge*, London: Sage.

Latour, B. (1990) "Drawing Things Together," in M. Lynch and St Woolgar (eds) *Representation in Scientific Practice*, Cambridge: The MIT Press, 19–68.

Toulmin, St (1953) *The Philosophy of Science: An Introduction*, London: Hutchinson's University Library.

Part I
Molecularizing maps

1 Mapping genes in microorganisms

Angela N.H. Creager

When T.H. Morgan's group developed the linkage analysis methods that enabled them to map genes on the chromosomes of *Drosophila melanogaster*, their methodology took advantage of the fact that the fruit fly is diploid and reproduces sexually. The recombination events that enabled clever scientists to locate genes were associated with meiosis; *breeding* was crucial to establishing linkage. Yet shortly after the Second World War, evidence surfaced for recombination in bacteria and viruses; a year later Joshua Lederberg published the first linkage map for *Escherichia coli* (Delbrück and Bailey 1946; Lederberg 1947). In 1949, Alfred Hershey and Raquel Rotman extended genetic mapping further down the microbial scale, producing the first linkage map of a virus, bacteriophage T2 (Hershey and Rotman 1949). What did it mean to extend the "mapping" of genes from sexually reproducing organisms such as flies and plants to viruses and bacteria, for whom even the existence of chromosomes was uncertain and disputed?[1] And how did genetic mapping in the microbial world give new meanings to *sex* and *mating*, and even to the term *gene* itself?

The view that microbes are authentic genetic organisms was far from secure when the first bacterial gene was mapped. As George Beadle had stated in 1945:

> The genetic definition of a gene implies sexual reproduction. It is only through segregation and recombination of genes during meiosis and fusion of gametes that the gene exhibits its unitary property. In bacteria, for example, in which cell reproduction is vegetative, there are presumably units functionally homologous with the genes of higher organisms, but there is no means by which these can be identified by the techniques of classical genetics.
>
> (Beadle 1945a: 18)

These problems did not deter researchers from trying to apply Mendelian and Morganian principles to bacteria, even though their size and lack of apparent sexual differentiation made it "difficult to distinguish between somatic and germinal elements, between character and factor, between phenotype and genotype" (Sapp 1990: 58). The 1943 fluctuation test by Salvador Luria and Max Delbrück (Luria and Delbrück 1943), showing that bacteriophage resistance in *E. coli* was attributable to mutation, and Lederberg and Tatum's 1946 observation of genetic recombination in *E. coli* (Lederberg and Tatum 1946b), provided bacterial geneticists with inspiration—and justification—for developing a thoroughly genetic approach to microbiology. The subsequent mapping of bacterial genes aimed at providing further demonstration that bacteria (as well as viruses) were amenable to genetic analysis. However, in their campaign to overturn what they viewed as widespread Lamarckianism among bacteriologists, bacterial geneticists tended to give short shrift to previous interpretations of bacterial heredity, variation, and even sexual exchange. Two bacteriological traditions were especially relevant to the postwar attempts to "geneticize" bacteria.

First, attention to the role of life cycles in microbes had already inspired bacteriologists to search for bacterial conjugation. In the 1910s and 1920s, growing interest in bacterial life cycles provided a framework for addressing morphological variation in bacteria without challenging their stability as biological species (Amsterdamska 1991). Sexual cycles had first been established in fungi, but these studies of fungal sexuality did not focus on genetics (e.g. the results of crosses) but on the mechanism of exchange, especially through cell fusion.[2] This morphological and microscopic orientation pervaded attempts to document bacterial sexuality before the Second World War, and played a part in postwar discussions of the role of life cycles in bacterial genetics.

Second, studies of enzyme adaptation in the 1930s provided a compelling framework for viewing the emergence of new traits in microbial cultures as physiological rather than genetic in nature. This interest in adaptation developed out of widespread nutritional studies of bacterial growth and fermentation. Biochemical interest in elucidating metabolic pathways was also part of broad-ranging nutrition research during the interwar period, and microbial variants with altered growth factor requirements were valuable experimental tools to biochemists. Nutritional research both provided the background motivating bacterial geneticists in the postwar period to contest non-hereditary explanations for microbial variation, and made available the experimental resources and materials for the first studies of microbial mutants and genetic recombination.

Much of this chapter will concern the "classic" experiments of bacterial genetics, which produced not only novel genetic maps but also an important new field of research. My rendition will highlight how these experiments drew on and contributed to developments in biochemistry and bacteriology, contexts that are sometimes overlooked when the events are invoked for their fundamental importance to molecular genetics. For example, Beadle and Edward Tatum's collaboration on the biochemical genetics of *Neurospora crassa* drew on the resources and techniques of thriving nutritional research programs—and contributed new methods for selecting microbial growth factor mutants. The expanding repertoire of nutritional mutants that followed in their wake accelerated the mapping of metabolic pathways in fungi and bacteria, which became materials of choice for intermediary metabolism and enzyme biochemistry (Singleton 2000).

The contributions from studies of microbial and viral genetics to debates in bacteriology were also significant. By the 1940s, longstanding disputes over the origin of bacterial variation nucleated around the problem of heredity—whether genetic change could account for variation. During the interwar period, as mentioned earlier, bacteriologists often attributed changing growth requirements of bacterial cultures to "adaptation" and "training" in response to environmental change. By contrast, bacterial geneticists argued that the ability of microbes to respond to environmental change could be understood in terms of rare genetic mutations, selected for by particular growth conditions. Nutritional mutants of *E. coli* contributed to an increasingly geneticized understanding of bacterial variation, one that drew adaptation into its explanatory framework.

A second and related concern among microbiologists was whether genes in bacteria were located on chromosomes analogous to those in the nuclei of the best-studied diploid organisms, or whether cytoplasmic agents of heredity might account for transmission or expression of bacterial traits. In the end, a chromosomal view of heredity prevailed in the microbial world, like that held by geneticists working on *Drosophila*, corn, and mice. Yet the unusual nature of bacterial genetics, including for *E. coli* the dependence of "mating" itself on cytoplasmic fertility factors, meant that the nucleocentrism was necessarily qualified.[3] The assimilation of cytoplasmic heredity into the framework of bacterial genetics had practical as well as conceptual consequences: autonomous genetic units such as fertility factors, viruses, and episomes became crucial tools for advancing the aims of gene mapping. Seymour Benzer employed bacteriophage T4 to take mapping down to a new level of precision—recombination distances of 0.0001%, at the scale of adjacent nucleotides (Holmes 2000 and Chapter 2, this volume). In the late 1960s, transduction provided a powerful technology for achieving higher resolution in bacterial genetic maps (Lederberg 1993). By this time, the continued mapping of alleles in bacteria was no longer aimed at stabilizing a genetic framework (which was already robust), but contributed to the technical infrastructure that made certain microbes the organisms of choice for research on cellular regulation, metabolism, protein synthesis, and a multitude of other current problems in biology.

Bacterial variation and nutritional mutants

The concerns of late-nineteenth-century bacteriology provide important background to twentieth-century debates about variation in microbes.[4] A belief in the biological stability of bacteria was implicit in Robert Koch's "postulates" linking specific bacterial pathogens to disease—as well as in his "pure culture" method.[5] Medical bacteriologists were not the first to view bacteria as discrete species; in 1875 botanist Ferdinand Cohn published his taxonomic system classifying bacteria morphologically and separating them from fungi and algae (Mazumdar 1995: ch. 2). This perspective, now labeled the Cohn–Koch theory, challenged a widespread bacteriological "polymorphism," informed by the manifold observations of microbes that gave rise to new forms, and exemplified by the fungi, whose "morphic cycles" were actively investigated by researchers (Gradmann 2000).[6]

Research on microbial life cycles was assimilated into the "monomorphic" understanding of the immutability of bacterial species.[7] Here the fungi provided a point of reference; sexual cycles and sporulation in fungi were well-studied by botanists in the late nineteenth century, and Alfred Blakeslee differentiated homothallic from heterothallic mating forms in 1904 (Blakeslee 1904).[8] Bacteria were labeled "fission fungi"; despite the emphasis on their vegetative reproduction, some bacteriologists believed they displayed fungal-like life cycles, including sexual cycles (Brock 1990: 30; Lederberg 1996). There was a taxonomic motive for documenting bacterial sexuality: bacteria were classified with plants, and identification of their sexual organs would provide a basis for Linnaean classification of the microbial world.[9] By the 1930s, the study of sexuality in certain other microbes flourished, with the identification of mating types and genetic segregation for the yeast *Saccharomyces* (Winge and Laustsen 1937), as well as Franz Moewus' work on the mating and biochemical genetics of the algae *Chlamydomonas* (Sapp 1990).

In the end, this early interest in microbial and even bacterial sexual cycles was not the route through which bacterial genetics emerged, as others have noted (Brock 1990: 31; Amsterdamska 1991). In part, this was due to the largely morphological orientation of studies of life cycles, including the sexual cycle—with its particular manifestation in cellular conjugation. Those bacteriologists who investigated putative sexual exchange in bacteria followed the example of fungi by focusing on the *mechanism* for genetic exchange, rather than its effects, for example, genetic linkage. Some researchers persisted through the 1940s in their efforts to visualize cell fusion events microscopically as evidence for bacterial sexuality.[10] More generally, however, as interest in life cycle ("cyclogenic") theories of microbes waned, so did interest in the significance of bacterial sexual cycles. By the 1930s, the attractiveness of attributing variation in bacteria to life cycle changes had declined as compared with the strong support of this interpretation in the literature two decades earlier (Amsterdamska 1991).[11]

Not that interest in bacterial variation declined; rather, bacteriologists had shifted focus from morphological to physiological variation as they adopted synthetic culture media to study microbial growth. The focus on bacterial growth and nutrition grew out of laboratory practices in agricultural bacteriology while reflecting the long history of research on fermentation.[12] In 1900, F. Dienert had shown that yeast grown in glucose could not immediately ferment galactose, but if cells were transferred to a galactose-containing medium that also had a nitrogen source, they could *acquire* the ability to grow on galactose (Dienert 1900; Brock 1990: 267). Conversely, cells that could grow on galactose, if transferred back to glucose-containing media, lost their ability to ferment galactose. Bacteria seemed able to respond to the presence of certain foodstuffs by generating *de novo* enzymes that could metabolize them, which Henning Karström dubbed "adaptive" enzymes in 1937 (Karström 1937).

During the 1930s, the prevalence of nutritional studies in bacteriology provided the context for ongoing interest in bacterial variation and adaptation.[13] Routine bacteriological investigation of a particular organism came to include studying its growth in "chemically definable media," along with analysis of the "growth factors," which were essential nutrients for the bacterium (Snell 1951: 222ff; Knight 1971: 16). The availability of all twenty purified essential amino acids by 1930, and many purified vitamins over the course of the 1930s, fed into these investigations of microbial nutritional requirements.[14] In addition, the use of pure culture technique enabled bacteriologists to isolate and analyze spontaneous variants with new growth factor requirements. The origin of these variants remained disputed. As André Lwoff assessed in 1946, "very few cases of such variations have been studied in such a way as to allow a conclusion regarding their *adaptive* or *mutative* nature" (Lwoff 1946: 150, emphases in original). Many bacteriologists reported being able to change the nutritional requirements of bacterial variant strains by changing media conditions: the scientist could "train" an auxotrophic strain (one that required a particular amino acid or nutrient) to grow without the substance by subjecting the culture to a decreasing concentration of it.[15] Such studies were often interpreted by bacteriologists as evidence that bacterial traits were not determined by genes, but elicited by environmental change.

The availability of agents that inhibited bacterial growth—especially antibiotics and bacteriophages—meant that variants resistant to these agents could be selected for and analyzed in the laboratory. As in the case of nutritional variants, there were debates about whether bacterial resistance (especially to antibiotics) was genetic or induced (the term adaptation was frequently used in this connection).[16] Michael Doudoroff interpreted his observations of the growth of bacteria in normally inhibitory concentrations of sodium chloride in terms of the principle of adaptation (Doudoroff 1940). On the other hand, in favor of a genetic interpretation, variable traits seemed to be linked in some bacteria. E.H. Anderson found that out of 58 phage-resistant mutant strains of *E. coli*, 27 also exhibited growth-factor deficiencies (Anderson 1944, 1946). Beyond the arena of nutritional research, the continuing debates over bacterial variation were kept alive by their medical relevance. Morphological and serological variability in bacteria had long complicated efforts to classify pathogens and establish bacteriological etiologies of disease (Amsterdamska 1987), and the widespread observations of antibiotic resistance in the 1940s posed a new challenge, both scientific and pragmatic in nature, to bacteriologists (Creager, unpublished).

George Beadle and Edward Tatum's pathbreaking work on *Neurospora crassa* took advantage of the pure culture strategies that were popular among bacteriologists during the interwar period. As others have noted, the choice of pink bread mold as a subject was highly pragmatic in nature, and reflected their previous difficulties with using *Drosophila melanogaster* as an instrument of biochemical genetics (Kay 1989; Kohler 1991, 1994: ch. 7). In 1935, Beadle and Boris Ephrussi had developed a method of tissue transplantation into larvae as a way to study physiological genetics, and their painstaking study of eye transplants led them to propose the sequence of pigment reactions responsible for eye color development (Beadle and Ephrussi 1936, 1937; Ephrussi and Beadle 1937).[17] Tatum began collaborating with Beadle to advance the chemical side of the project, by extracting the pigment precursors from *Drosophila* larvae for characterization. In 1941, he was endeavoring to isolate the v^+ eye "hormone" when the German chemists Adolf Butenandt and Alfred Kühn beat Tatum to identifying the chemical substance as a tryptophan derivative (Lederberg 1960; Kohler 1991: 114).

As Kohler has emphasized (1991), *Drosophila* made a relatively poor chemical instrument, especially as compared with its productiveness for chromosomal mapping. When Beadle and Tatum turned to *Neurospora*, the basic genetics had already been worked out, and importantly, the fungus could be grown on chemically defined media, making it amenable to biochemical investigation (Beadle 1974).[18] Indeed, Tatum had drawn on his extensive background in nutritional research (even insect nutrition; he had studied the starvation effect in flies as well as determined nutritional requirements for flies and a variety of microbes) in trying to advance the biochemical analysis of larval transplants. But *Neurospora* was much more amenable to nutritional experimentation.[19] Its growth requirements were quickly established, and x-rays could be used to generate mutants from spores (which was even simpler than the procedure for *Drosophila*).

Beadle and Tatum mutagenized *Neurospora* spores with x-rays, crossed these potential mutants with wild-type spores, separated out the resulting ascospores, and grew these up as haploid cultures in rich media.[20] Cultures were screened by transferring bits of each colony (i.e. conidia) to test tubes containing minimal media (sugar, salts including a nitrogen source, and biotin). Mutants with new nutritional requirements—presumably on account of a disrupted metabolic step—would not grow. The researchers could then pinpoint the deficiency by putting these colonies through a battery of nutritional tests in minimal media supplemented with specific amino acids and vitamins.

By the fall of 1941, this method had yielded three good mutants with deficiencies in synthesizing vitamin B_1 (thiamine), vitamin B_6 (pyridoxine), and *para*-aminobenzoic acid (Beadle and Tatum 1941; Tatum and Beadle 1942; Kay 1989: 83). The first growth factor requirement observed (pyridoxine) segregated in crosses according to Mendelian expectations, suggesting that it was due to a mutation at a single genetic locus. The steady appearance of nutritional mutants continued as thousands of cultures were screened, and soon Beadle and Tatum had more mutant strains than they could follow up. The work and the research group expanded rapidly, despite the mobilization of scientists in the American war effort. As Lily Kay has pointed out, the potential use of their *Neurospora* strains in bioassays of vitamins and amino acids garnered strong financial support from the Office of Scientific Research and Development's Committee on Medical Research and from pharmaceutical houses—as well as from patrons, such as the Rockefeller Foundation, interested in the "pure" research aims (Kay 1989).

In the end, the payoff was not so much genetic as biochemical: the mutants provided an unprecedented resource for mapping metabolic pathways for the biosynthesis of vitamins and amino acids (Kohler 1991: 121–4).[21] It is worth noting that the ambition to use microbes to investigate metabolism predated the research on *Neurospora*. As Susan Spath has pointed out, Beadle and Tatum's methodology of obtaining genetic mutants blocked at specific points in biochemical pathways resembles A.J. Kluyver's studies of naturally occurring microbial mutants and C.B. van Niel's elective culture technique (Spath 1999: ch. 3). Yet Beadle and Tatum's program for *generating* as well as analyzing mutants gave existing interest in using genetic methods to study metabolism a more powerful

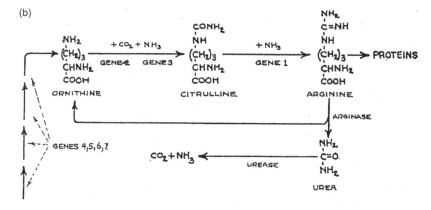

Strain number	Growth on				
	Arginine	Citrulline	Ornithine	Proline	Glutamic acid
3,485	+	−	−	−	−
6,572	+	+	(±)	−	−
6,155	+	+	+	−	−
9,929	+	+	+	+	−
24,033	+	+	+	+	+

Figure 1.1 Use of mutant strains to elucidate the ornithine cycle in *Neurospora*, which includes biosynthesis of arginine. Results were based on analysis of fifteen "arginineless" mutant strains (requiring arginine to grow), each of which represented a specific genetic block in the biosynthesis. (a) Growth of various arginineless strains on minimal media plus an amino acid or intermediate, from Bonner, D. (1946) "Biochemical Mutants in *Neurospora*," *Cold Spring Harbor Symposia on Quantitative Biology*, 11: 19. (b) Diagram of ornithine cycle showing assignment of specific genes for enzyme-catalyzed steps in the pathway based on mutant growth patterns. Reproduced with permission from Beadle, G.W. (1945) "Genetic and Metabolism in *Neurospora*," *Physiological Reviews*, 25: 654.

experimental foundation. By virtue of being able to generate many mutant strains with the same nutritional deficiency, one could use supplementation, with compounds thought to be intermediates in metabolic pathways of the growth factor, in order to determine exactly where in the pathway a particular strain was blocked. As Beadle illustrated for the laboratory's work on "arginineless" mutants, the growth of the strains on various intermediates enabled his group to represent the sequence of enzyme-catalyzed steps in the ornithine cycle, with genes assigned to each observed metabolic block (Beadle 1945b) (Figure 1.1).

While the Beadle–Tatum experiments are usually invoked for their theoretical significance to genetics, namely demonstrating the "one gene-one enzyme" correlation, the material consequences of their work for postwar microbiology and biochemistry—the use of genetic mutants to mark out biochemical pathways—were just as profound, and less contested.[22] Most notably, even though Beadle and Tatum did not map genes in *Neurospora*, their method associated nutritional mutants with the representational activities of mapping, namely the maps of metabolic pathways.

From nutritional mutants to genetic mapping in *E. coli* and bacteriophage

Tatum soon began a systematic search for similar growth-factor-requiring mutants in *Escherichia coli* strain K-12 and in *Acetobacter melanogenum*, using the same general approach that had worked for screening mutagenized *Neurospora*. In 1944, he reported on two growth-requirement-deficiency strains of *E. coli* out of 800 colonies screened, one requiring biotin and the other requiring threonine (Gray and Tatum 1944).[23] He had also obtained four mutant strains of *Acetobacter melanogenum* (from 575 cultures tested), three that required amino acids and one that required the purine adenine. Such supplement-requiring mutants (termed auxotrophs in 1950[24]) held promise for elucidating the bacterial genes responsible for the synthesis of growth factors. Nonetheless, Tatum was cautious in his interpretation:

> Definite proof of the existence of genes in bacteria cannot be obtained by the methods of classical genetics since no sexual phases in their life cycles have yet been demonstrated. For the same reason it is at present impossible to distinguish between mutations in nuclear genes and in cytoplasmic determiners or plasmagenes in bacteria. The term "gene" can therefore be used in connection with bacteria only in a general sense. Nevertheless, the role of genes in bacterial variation has been suggested and the existence in bacteria of discrete nuclear material, perhaps analogous to chromosomes, has been demonstrated in certain forms.
>
> (Gray and Tatum 1944: 408)

Joshua Lederberg, working in Tatum's laboratory (now at Yale), continued this line of investigation. Lederberg and Tatum developed methods for more easily detecting biochemical mutants, growing colonies first on minimal media (on which the wild-type cells would grow), then adding a multiple supplement enabling small colonies of growth-requirement mutants to appear (Lederberg and Tatum 1946a). These smaller colonies could then be picked for further isolation and nutritional screening. Tatum had already constructed some double-nutritional-mutant strains of *E. coli*; Lederberg added to this set of multiple growth requirement deficiencies and put them to use in looking for genetic exchange.[25] This venture built on Lederberg's earlier attempts, in the spring and summer of 1945 in Francis Ryan's Columbia laboratory, to detect genetic exchange mediated by DNA in *Neurospora* mutant strains.[26] However, the rate of reversion of single mutants of *Neurospora* had been too high to be able to effectively screen for new (non-revertent) genotypes.[27] The availability of multiply-deficient strains in *E. coli* helped alleviate these technical artifacts associated with reversion; spontaneous double (or triple) reversions would be negligible.[28]

When Lederberg and Tatum grew multiple mutants in mixed cultures (which can be generically represented in terms of the genetics as $A^-B^-C^+D^+$ mixed with $A^+B^+C^-D^-$) on minimal media, they found that prototrophic colonies appeared—that is, colonies that had recovered the ability to synthesize the growth factors they previously required (i.e. $A^+B^+C^+D^+$).[29] Reversion could not account for the frequency with which these wild-type-like strains appeared (10^{-7}); these prototrophic colonies from mixed cultures represented novel genotypes, presumably generated through genetic exchange.

Lederberg and Tatum first presented this evidence for mating in *E. coli* at the 1946 Cold Spring Harbor Meeting on "Heredity and Variation in Micro-organisms" (Lederberg and Tatum 1946b).[30] Further studies in 1946 emboldened Lederberg and Tatum to publish a note in *Nature* entitled "Gene Recombination in *Escherichia coli*" (Lederberg and Tatum 1946c). Using triple mutant strains (Y-10, requiring threonine, leucine, and thiamine, and Y-24, requiring biotin, phenylala-nine, and cystine), Lederberg and Tatum recovered not only complete prototrophs, but an assortment of single and double deficiencies (e.g. strains requiring biotin and leucine, biotin and threonine, and biotin and phenyalanine). They concluded that either sexual fusion and recombination had occurred, or transforming factors had transmitted the genes for growth factors from one strain to another. Either way, "[t]hese experiments imply the occurrence of a sexual process in the bacterium *Escherichia coli*" (Lederberg and Tatum 1946c) (Figure 1.2).

The 1946 Cold Spring Harbor meeting where Tatum and Lederberg announced their initial results proved to be a propitious meeting for microbial genetics. At the same meeting, Alfred Hershey presented evidence that bacterio-phage mutant strains could undergo genetic recombination.[31] At the outset of his presentation, he was careful to attend to the differences between observations of viral genetics and classical genetics. In particular, he made clear that viral genes were defined by *mutation*, not by recombination[32]:

> Owing to the nature of the material, our principal tools for the study of viral genetics are the analysis of mutational patterns and the measurement of mutation rates. For this reason alone, viral genetics is bound to differ in super-ficial appearance from classical genetics, where the principal tool is segrega-tion. When we speak of independent genetic sites in the viral particle, we will mean loci that mutate independently of other loci. Their relation to the genetic locus recognized as a crossover unit is dubious to say the least, though this cannot be said to diminish their importance. But it must be kept in mind that genetic factors recognized to be independent by analysis of mutations in a virus, may prove to be structurally analogous to factors classified as alleles by segregation experiments with other materials.
>
> (Hershey 1946: 67)

Even with these caveats, he argued that experiments showing genetic modifica-tion of viruses offered "an equivalent of, or a substitute for, meiotic segregation in

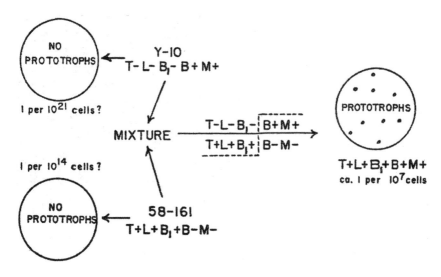

Figure 1.2 Schematic diagram representing recombination experiment producing prototrophs from two multiply-deficient strains of *E. coli* K-12 (Y-10 and 58-161). Each circle represents a minimal media agar plate containing the strain(s) indicated, with prototrophic colonies represented as small dots. The expected frequency of prototrophic genetic revertants, which were not actually observed, on the plates at left (e.g. 1 per 10^{21} cells?) is contrasted with the higher observed incidence of prototrophy from the mixed strains on the plate at right (*c.* 1 per 10^7 cells), which is attributable to recombination. The letters represent particular growth requirements: T for threonine, L for leucine, B_1 for thiamine, B for biotin, and M for methionine. Reproduced with permission from Lederberg, J. and Tatum, E.L. (1953) "Sex in Bacteria: Genetic Studies, 1945–52," *Science,* 118: 170. Copyright: American Association for the Advancement of Science (1953).

the genetic analysis of viruses" (Hershey 1946: 67). He proceeded to revamp the definitions of mutation, genetic site, allele, and genetic complex to be consistent with observations from bacteriophage strain research. By mutation, Hershey meant a "change in any property of a virus which occurs in a single step from one recognizable step to another." Building on this definition, genetic site was to refer to "the hypothetical structural site of a mutation or set of alternative (allelic) mutations that occur independently of other known mutations. When the word locus is used in connection with viral genetics, it is necessarily the genetic site that is meant" (Hershey 1946: 67). Thus gene was defined in such a way as to make new alleles mappable without reference to nucleus or chromosome as the site of the structure.

Working with various host-range mutants (denoted by *h*) and lysis-inhibition mutants (denoted by *r*) of phage T2, Hershey showed that when bacterial cells were infected with two types of phage, novel viral phenotypes appeared.[33] Specifically, from a mixed phage infection with h^+r clones and $h^rh^br^+$ clones, Hershey obtained seven out of eight possible phenotypes that could result from recombination if these traits were carried by independent alleles. The apparent segregation of these genetic factors was highly suggestive of either crossing over or "actual exchange of genetic material," perhaps like that associated with the transformation of pneumococcal types (Hershey 1946: 75).

At the same Cold Spring Harbor meeting, Max Delbrück and William Bailey also presented evidence that bacteriophage mixed-infection resulted in novel viral genotypes. Following up earlier observations of mutual exclusion—the fact that in mixed infection with two different bacteriophages, one is excluded from multiplication—they tried mixed infections with variants of the *same* bacteriophage or closely related phages. Some results were surprising. Mixed infections with T6r and T4r^+, for example, yielded some virus particles with T6r^+ and some with T4r phenotypes. Rather than interpreting this as recombination, as Hershey had, Delbrück and Bailey argued that the presence of one mutant *induced* the production of a similar mutation in the other strain (Delbrück and Bailey 1946). In fact, this was one common explanation for the transforming factor result in *Pneumococcus*—that the agent induced the mutation responsible for the rough-to-smooth transformation in the recipient strain.[34]

Thus, many results of 1946 pointed to genetic exchange between viral variants and between *E. coli* strains, opening up the prospect of genetic mapping. The mechanism of genetic recombination in *E. coli* was unclear—there was no evidence for a diploid state or heterokaryon. Similarly, the mode of viral gene recombination was unknown and Hershey had left open the possibility that genetic exchange between T2 phage strains was more akin to transformation than crossing over during meiosis. Yet despite these uncertainties it seemed plausible that bacterial and viral genes might be mapped, which would further bolster the notion that bacteria and viruses inherited traits in fundamentally the same way as diploid organisms. Lederberg immediately pursued this goal, although he had to be creative in adapting his experimental setup to develop a way to quantify linkage.

The power of Lederberg's prototroph recovery technique was that it allowed for the selection of rare recombinants. Yet the fact that non-prototroph recombinants were not recovered meant that the class of segregants available for analysis was incomplete. In order to circumvent this problem, Lederberg employed non-selected traits such as resistance to bacteriophage and lactose fermentation to see how these traits segregated. Recombinants were obtained with these non-selected traits, and they showed regular patterns of genetic segregation. For example, a mixture of B$^-$M$^-$P$^+$T$^+$R and B$^+$M$^+$P$^-$T$^-$ (where B represents ability to synthesize biotin, M represents ability to synthesize methionine, P represents ability to synthesize proline, T represents ability to synthesize threonine, and R signifies resistance to phage T1) generated ten isolated prototrophs, eight of which were resistant to phage, and two of which were sensitive (Lederberg and Tatum 1946b). The fact that the (likely recessive) trait was observed in the "f-1" generation confirmed that *E. coli* is haploid.[35] When the cross was reversed, the segregation pattern was roughly inverted, suggesting that phage resistance was chromosomally linked to the selected nutritional markers. As Lederberg and his co-workers later put it, "the occurrence of a given allele among the prototrophs is regulated not by

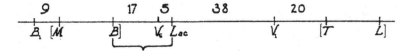

Figure 1.3 The first linkage map for *E. coli* K-12, based on markers for synthesis of growth
requirements, growth on lactose, and bacteriophage resistance (B_1 is for thia-
mine, M is for methionine, B is for biotin, V_6 is resistance to bacteriophage T6,
Lac is ability to grow on lactose, V_1 is resistance to bacteriophages T1 and T5,
T is for threonine, and L is for leucine). Reprinted with permission from
Lederberg, J. (1947) "Gene Recombination and Linked Segregations in
Escherichia coli," *Genetics*, 32: 516.

the nature of the allele, but by its parental coupling" (Lederberg *et al.* 1951: 416).
The addition of particular growth factors to the plating medium allowed for treat-
ment of individual nutritional loci as unselected markers, so that the order of
linked nutritional factors could also be established.

Lederberg used the consistent segregation ratios he obtained from the crosses
of Tatum's double-mutant strains to deduce the first genetic map of *E. coli*, which
he published in 1947 (Lederberg 1947) (Figure 1.3). The paper postulated two
linkage groups associated with a single chromosome, one group containing the genes
for synthesizing thiamine, methionine, and biotin, as well as the gene for lactose fermentation, and the other containing genes for phage resistance and for the synthesis
of threonine and leucine. Lederberg and Tatum favored viewing recombination "as
a consequence of cell fusion, 'karyogamy' and meiosis with crossing over," although
they could not at this time rule out the transmitting of genes via transforming-factor-
like agents (Lederberg 1947: 520).[36]

Hershey made similarly rapid progress studying recombination in bacteriophage T2. He (and coworker Raquel Rotman) first established linkage using virus
mutants exhibiting differences in lysis-inhibition (r^+) and rapid lysis (r). "Crosses"
of these mutant strains—that is, mixed infection—yielded "offspring" whose phenotypes could be genetically analyzed (Hershey and Rotman 1948). The distribution of phenotypes among progeny was used to infer linkage, that is, a low rate of
recombination would be expected if traits were linked genetically. Extending the
analysis to include host-range and plaque-type mutants, Hershey and Rotman
were able to construct a linkage map of the bacteriophage genes (Hershey and
Rotman 1949) (Figure 1.4).

As Thomas Brock has astutely noted, the mapping of bacteriophage genes fundamentally altered the language and representations used by phage geneticists.
Phrases such as "mutual exclusion" that had been coined to describe phenomena
that seemed specific to phage were soon replaced by the classical genetic terminology of "loci" and "linkage" (Brock 1990: 137). Moreover, the phage could no longer
be considered as (or analogous to) a single gene; it contained a genome, if small.
The implicit reference system was now *Drosophila*, the exemplar of chromosomal
mapping (Kohler 1994). This shift in phage work towards the conceptualization of

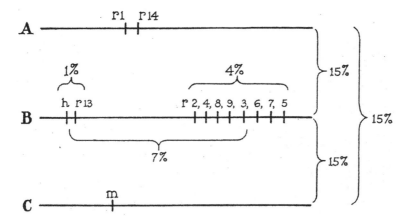

Figure 1.4 Schematic diagram depicting linkage relations among plaque-size (m), rapidly-lysing (r) and host-range (h) mutants of phage *T2H*. The percentages indicate yields of wild type in two factor crosses. Diagram and caption from Hershey, A.D. and Rotman, R. (1949) "Genetic Recombination between Host-Range and Plaque-Type Mutants in Single Bacterial Cells," *Genetics*, 34: 49. Reproduced by permission of the Genetics Society of America.

the gene as an *instrumental* entity (to draw on Raphael Falk's terminology; Falk 1986) was reinforced through the intensive work on recombination with T phage strains in the 1950s (e.g. Doermann *et al.* 1955).[37] This was not Max Delbrück's original intention in developing an experimental system with bacteriophage—he had hoped to access the gene as a *material* entity, the physico-chemical gene at the heart of H.J. Muller's research program. Instead, phage was made into a mapping system *qua* Morgan.

Methods for isolating mutants in *E. coli* were rapidly improved through use of selective agents such as penicillin (Davis 1948; Lederberg and Zinder 1948). Bernard Davis' observation that cell-to-cell contact was necessary for mating strengthened the hypothesis that the mating interaction consisted of cell fusion followed by some sort of fertilization–reduction cycle (Davis 1950).[38] Tracy Sonneborn's study of sexual conjugation in *Paramecium*, in which mating cells exchange haploid nuclei across a cytoplasmic bridge, provided the best-studied example suggesting how *E. coli* mating might take place.[39]

Yet for bacteria, mating remained frustratingly rare. The low frequency of pro-totroph recombinants meant that one had little hope of isolating a conjugating bacterial pair to analyze.[40] In addition, it remained unclear how widespread mating was in bacteria like *E. coli*. While other researchers were able to repli-cate Lederberg and Tatum's mating experiments with auxotrophs of K-12 (Cavalli-Sforza 1950; Newcombe and Nyholm 1950), many laboratory strains of *E. coli* still failed to exhibit any genetic exchange. Luca Cavalli and H. Heslot, working in R.A. Fisher's laboratory at Cambridge, showed that only one of seven

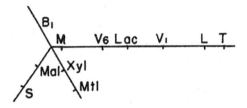

Figure 1.5 Schematic representation of linkage data from crossing *E. coli* K-12 derivative
strains 58-161 × W-1177. Letters refer to genetic loci associated with synthesis
of growth requirements, fermentation of sugars, and drug resistance as follows:
B_1 is for thiamine, M is for methionine, V_6 is resistance to bacteriophage T6, Lac
is for growth on lactose, V_1 is resistance to bacteriophages T1 and T5, L is for
leucine, T is for threonine, Mal is for growth on maltose, Xyl is ability to grow
on d-xylose, Mtl is ability to grown on mannitol, and S is for resistance to streptomycin. The original caption states: "This diagram is purely formal and does
not imply a true branched chromosome." Reproduced from Lederberg, J.,
Lederberg, E.M., Zinder, N. and Lively, E.R. (1951) "Recombination Analysis of
Bacterial Heredity," *Cold Spring Harbor Symposia on Quantitative Biology*, 16: 417.

E. coli strains in the British culture collection was capable of mating with one of
Lederberg's test strains (Cavalli and Heslot 1949).[41] In Lederberg's own canvass-
ing of 140 clinical isolates of *E. coli*, only nine were capable of mating with strain
K-12 (Lederberg 1951a; Brock 1990: 88).[42]

More troubling was the fact that the growing body of linkage data from K-12
auxotrophs could not be made consistent with the hypothesis of chromosomal
linearity, which fit so well with early evidence. In an important summation of the
recombination data at the 1951 Cold Spring Harbor meeting, Lederberg and his
colleagues presented a four-armed diagram, although they demurred at hypothe-
sizing that the bacterial chromosome was actually branched (Lederberg *et al.*
1951: 417)[43] (Figure 1.5). In this way, the mapped bacterial genes were truly
instrumental entities (in Falk's sense) whose relation to physico-chemical determinants of heredity remained unclear. In a 1953 review, Lederberg and Tatum
speculated that *E. coli* contained a second chromosome, yet one that did not seg-
regate entirely independently of the first (citing Fried and Lederberg 1952,
Lederberg and Tatum 1953: 172).

The inscrutability of *E. coli*'s mating habits was the basis for a biologist's ditty,
"dedicated" to Joshua Lederberg and sent to him in 1950 by Bernard Davis:

> Confess, K-12, do you conjugate,
> When fanned by flames erotic?
> Can we believe you really mate?
> Is fission then meiotic?
>
> You long eluded watchful eye
> When cultures were not mixed.

Pray tell, dear lady, why so shy?
Once in ten to the sixth!

The evidence is indirect,
Lysenko would have a fit!
"On casts of dice you bourgeois build
Your theories, full of wit."

Perhaps it's done with mirrors
Or with principles dissolved.
Yet round a union much more firm
The world has e'er revolved.

...

With N-Z for muscle and sugar for heat,
With phosphate to make you fertile,
With potash to give your flagella a flip,
Please—take off that chastity girdle.

So mate on the plate or hide in its dew;
Or sleep in the deep or lie out in full view;
Or fuse to amuse or to have babies too—
But somehow unite and recombine—do![44]

The figuring of *E. coli* as a coy mistress enabled lively scientific banter, but the metaphor was also conceptually important. In particular, the representation of microbial genetic exchange as (heterosexual) "sex" informed developments in bacterial genetics in the "post-Hayes" era.[45]

Bacterial sexual differentiation—fertility factors, conjugation, and transduction

In the early 1950s bacteriologist William Hayes set out to study the kinetics of the bacterial mating process, and his results helped illuminate the basis for the peculiar mating patterns observed with *E. coli*. Hayes obtained basic K-12 strains from Cavalli, whom he met at Cambridge. Using basic auxotrophic strains in which streptomycin resistance had been selected, Hayes investigated how the addition of streptomycin at various time points during mating and growth of prototrophs affected the frequency of recombination. Much to his surprise, in one cross of a streptomycin-sensitive strain with a streptomycin-resistant strain, he found that prototrophs grew irrespective of when streptomycin was added. Even if streptomycin was present before mating, such that it presumably killed cells from the sensitive strain, recombination yielding prototrophs occurred anyway.

The asymmetry of mating—the fact that one strain "produced prototroph colonies on every occasion," whereas the same combination of traits but with the parent strains reversed failed to produce recombinants—suggested that genetic "exchange" was actually unidirectional (Hayes 1952a: 118). He initially thought of it as a type of gene donation, speculating that recombination occurred via virus-like genetic elements that could transfer the genes even after the host cell was killed. In a further publication (1952b), Hayes suggested this virus-like gamete might be related to the lysogenic phage λ (of which more is discussed later).[46]

Further studies by the Lederbergs and Cavalli (Lederberg *et al.* 1952; Cavalli *et al.* 1953) and by Hayes (1953) implicated an extrachromosomal factor, not one of the lysogenic phages, but a new agent termed the "F" factor (for fertility). This entity, which behaved as an infectious particle, was carried by some strains (termed F$^+$, or "male") and absent in others (F$^-$ or "female"). The F factor had the ability to replicate independently of the chromosome, but when it was transferred between cells, it could carry bacterial genes with it. The heterosexual language was as misleading as it was useful: although males do "give" sperm to females in copulation, fertilization reflects equal genetic contributions from both partners in the form of their gametes. What the language of fertility for bacteria reflected was not biological sex, or even anisogamy, but rather gender stereotypes—specifically the coding of males as active donors and females as passive recipients.[47]

At another level, acceptance of these findings consigned American bacterial geneticists to concede the importance of cytoplasmic inheritance, despite its Lamarckian associations. Interest in—and controversy about—non-nuclear determinants of heredity in the 1940s provided an important scientific context for early microbial genetics, since many of the best-studied examples of the "inheritance of acquired characteristics" were associated with bacteria (Sapp 1987). In the early 1940s, Sewell Wright and C.D. Darlington advanced the term "plasmagene" to refer to a self-reproducing cytoplasmic genetic unit that exhibited non-Mendelian inheritance (Wright 1941: 501; Darlington 1944).[48] Both noted the similarities between viruses and plasmagenes, while Wright also indicated how plasmagenes might provide a link between nuclear genes and cytoplasmic enzymes, which could account for well-documented phenomena of physiological change in bacteria (Wright 1945).[49] Along these lines, Sol Spiegelman elaborated a plasmagene-based model for enzyme adaptation in the late 1940s (Spiegelman and Kamen 1946; Brock 1990: 267–73; Gaudillière 1992). The unclear relationship between nuclear genes and plasmagenes raised thorny questions—were plasmagenes derived from nuclear genes or did they have an independent existence (like plant plastids)? Viruses would seem to be an example of genetic autonomy, although lysogeny in bacteriophages complicated the picture, with latent viruses apparently incorporated into normally dividing cells. Many mainline geneticists were uncomfortable with treating non-Mendelian inheritance as authentic and genetically significant, with H.J. Muller a prominent critic of plasmagenes (e.g. Muller 1951; see Sapp 1987: 117–18).

Debates over the generality and significance of cytoplasmic inheritance were further polarized in the late 1940s by the Lysenko affair (Sapp 1987: ch. 6). Proponents of Lysenkoism regularly cited work on cytoplasmic and infective heredity— such as Tracy Sonneborn's demonstration that antigenic types in *Paramecium* could be environmentally altered or Oswald Avery, Colin MacLeod, and Maclyn McCarty's isolation of the transforming factor for *Pneumococcus*—as evidence of the inadequacy of Mendelian genetics to account for new traits. Bacterial geneticists thus found themselves fighting on two fronts: during the same years that they sought to combat Lamarckian views of microbial heredity among bacteriologists (in an effort to establish that bacteria could be studied using Mendelian genetics), prominent bacterial geneticists such as Salvador Luria and Milislav Demerec were active alongside Theodosius Dobzhansky, Curt Stern, L.C. Dunn, and H.J. Muller in a campaign by Western scientists to support Russian geneticists by discrediting Lysenkoism (Krementsov 1996: 242–4). Thus it was ironic that the very capacity of *E. coli* to exhibit mating, the trademark of Mendelian organisms, relied on an entity that displayed cytoplasmic inheritance. It was clear by the early 1950s that non-nuclear heredity could not be relegated to the fringes of bacterial genetics.

It is notable that among bacterial geneticists, Joshua Lederberg had been arguing that struggles against Lamarckianism should not be used to ignore the reality of non-nuclear hereditary units, which appeared so consequential for bacterial genetics.[50] He introduced the term "plasmid" in 1952 to serve "as a generic term for any extra-chromosomal hereditary determinant" (Lederberg 1952: 413; see also Sapp 1987: 122). Observations in microbial genetics that had previously appeared to be the "inheritance of acquired characteristics" might now be accounted for through the transmission of acquired, autonomous hereditary units, or, as Peter Medawar had phrased it, "the casual leakage of self-reproducing particles from one cell into neighbors" (Medawar 1947: 371).

The discovery of transduction in *Salmonella typhimurium* by Norton Zinder and Lederberg in 1952 provided yet another example of gene transfer via an autonomous hereditary unit. Lederberg had set graduate student Zinder on the project of looking for genetic exchange in *Salmonella* based on the promising collection of serotypes, which might have plausibly arisen through recombination events (Zinder 2002). Zinder constructed multiply-deficient mutant strains like those Lederberg and Tatum had used in their *E. coli* mating experiments, and particular combinations of the *Salmonella* double mutants yielded prototrophs. Zinder and Lederberg analyzed the segregation of non-selected traits (e.g. ability to ferment specific sugars) in the prototrophs, and found only genes for selected traits were transferred. Moreover, unlike in the case of *E. coli* mating, cell-to-cell contact was not required for recombination in *Salmonella*—the selected traits were conferred by a "filterable activity" (FA). This agent was similar to the pneumococcal transforming principle in activity but not in composition—it was resistant to DNase, for instance.

Zinder and Lederberg connected their results to the ongoing bacteriological attempts to visualize mating morphologically, noting that FA was "reminiscent of the L-forms of bacteria," a putative filterable stage of the bacterial life cycle (Zinder and Lederberg 1952: 684).[51] However, the morphological results reported by the bacteriologists proved difficult to reproduce and their relationship to the genetics remained uncertain. A more significant association could be made between transduction and lysogenicity, because the donor *Salmonella* strain was also lysogenic for phage P22. This phage appeared to play some role in transduction, but the role of phage in the process of genetic transfer remained unclear. Zinder and Lederberg stressed in their 1952 paper that the bacteriophage particle could only serve as a passive carrier of the transduced material (Brock 1990: 194–6).

One of the best-accepted representatives of "infective heredity" was lysogenic phage λ. Esther Lederberg discovered in 1951 that *E. coli* K-12 contained this phage, which was named in analogy to the well-studied self-reproducing particle κ (kappa) in *Paramecium* (E.M. Lederberg 1951).[52] At that summer's Cold Spring Harbor meeting, the Lederberg laboratory presented λ as an exemplar of "extranuclear heredity" in *E. coli*, which was thought to reside and replicate in the cytoplasm during lysogeny (Lederberg *et al.* 1951: 435). Because of the well-developed genetics of K-12, there was an opportunity to use the techniques of mapping to locate λ—as a cytoplasmic particle, it would not be expected to exhibit linkage to chromosomal markers. However, the Lederbergs' mapping experiment in 1953 revealed that λ was genetically linked to a chromosomal marker associated with galactose fermentation (Lederberg and Lederberg 1953). As François Jacob and Elie Wollman commented,

> When first envisaged, this conclusion appeared somewhat surprising, since it would seem a priori that the noninfective structure which, in lysogenic bacteria, carries the genetic information of a virus, the bacteriophage, should be cytoplasmic rather than chromosomal. Although this problem was never seriously considered before 1950, that is until the investigations of Lwoff, the hypothesis of a cytoplasmic determination of lysogeny had been accepted implicitly.
>
> (Jacob and Wollman 1961: 90; also quoted in Brock 1990: 182)

Thus, experiments with λ revealed that the phage was part of the bacterial chromosome during lysogeny and became a self-reproducing genetic particle following induction—in other words, the partitioning of genes between the cytoplasm and the chromosome was not fixed. Further studies of bacterial sexuality reinforced this point.

At the Pasteur Institute, after years of careful work on lysogeny in *Bacillus megaterium*, Wollman and Jacob initiated work on phage λ in *E. coli*. The linkage between λ and the *E. coli* gal₄ gene provided a genetic handle for studying the phage during lysogeny—during its existence as prophage, as André Lwoff had termed it (Lwoff 1953). Working with high-frequency mating (Hfr) strains of *E. coli*, Wollman and Jacob found that recombination results for the *lambda* and *gal* loci depended on which parent contributed the phage, particularly in mating Hfr with F⁻ strains. The high frequency of lysis (by λ) when the Hfr strain was the lysogenic parent and the other parent was non-lysogenic was termed zygotic induction—the expected *lambda*⁺*gal*⁺ recombinants could not be recovered because lysis was induced.[53]

Soon zygotic induction became incorporated into new techniques of mapping. In crosses between F⁻ strains and different Hfr prophage-containing strains, one could determine the *order* of transfer of various markers by "interrupting" mating at various points.[54] Wollman and Jacob demonstrated that time of entry of the markers correlated with genetic linkage, such that one could use time points to establish a spatial map. At the 1956 Cold Spring Harbor meeting, Wollman, Jacob, and Hayes presented a genetic map of the Hfr H(λ) chromosome (Wollman *et al.* 1956). In this and other studies, each independently isolated Hfr strain gave the same order of genes, although the starting points ("origin") for transfer varied. As a consequence, Jacob and Wollman speculated that the chromosome of *E. coli* K-12 could be formally represented as a closed circle (Figure 1.6) (Jacob and Wollman 1961).

A circular genetic map, which accommodated evidence that *E. coli* genes form a single linkage group, was a novel representation. The topological problems

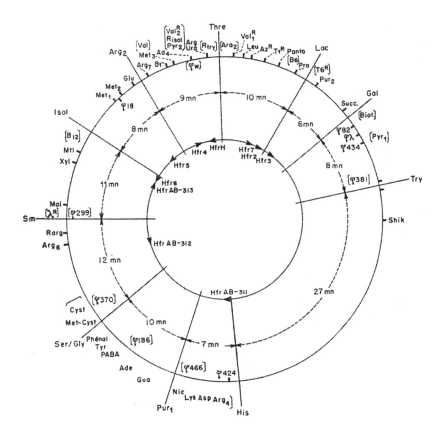

Figure 1.6 Schematic representation of the linkage group of *E. coli* K-12. The outer line represents the order of the characters (not their absolute difference). The dotted lines represent the time intervals of penetration between pairs of markers corresponding to the radial lines. The inner line represents the order of different Hfr types. Symbols correspond to loci for synthesis of various amino acids, vitamins, and other metabolites; fermentation of various sugars; resistance to bacteriophages, chemicals, and drugs; location of inducible phages (denoted by φ); and repression for arginine, isoleucine, and tryptophan. Diagram and caption reproduced by permission from Jacob, F. and Wollman, E.L. (1961) *Sexuality and the Genetics of Bacteria*, New York: Academic Press, 165.

presented by Watson and Crick's double-helical model of DNA had been extensively discussed (see Holmes 2001: ch. 1), and the "replication problem" was even more acute for a circular chromosome. As Brock has pointed out, Jacob and Wollman did not actually claim that the bacterial chromosome was circular—once again, it was the gene as an *instrumental* entity that they addressed (Brock 1990: 103). However, the round map soon began to take on physico-chemical meaning, as John Cairns published autoradiographs of tritium-labeled *E. coli* DNA that showed a circular chromosomal structure (Cairns 1963). Other aspects of bacterial genetic structure were coming into view as well. By the time Jacob and Wollman's monograph was published, electron microscopist Thomas Anderson contributed pictures of bacteria in the process of conjugation, which was no longer a hypothetical process. The fusion of cells during conjugation could be clearly seen (Jacob and Wollman 1961: 114–17), and the "zygotes" produced were now understood to be recipient cells with a set of transferred chromosomal genes from the donor, which could then recombine with the host chromosome.

Advances in the understanding of transduction also contributed to the high-resolution mapping of bacterial genes. By the mid-1950s, Zinder, Lederberg, and their coworkers had clarified the role of lysogenic phage P22 in transduction for *Salmonella*. Phage-mediated transduction relied on the fact that some phage particles carried only bacterial DNA, which could be transmitted into a recipient cell. Zinder emphasized the similarity of the process to transformation in *Pneumococcus*, although transduction did not involve naked DNA; a phage particle provided the vehicle of transmission (Zinder 1955; Brock 1990: 196). It was the infectious-like nature of transduction that Lederberg stressed: "the equivalence of virus and gene is here a formality of the means of hereditary transmission" (Lederberg 1956: 277). Transduction not only reinforced these conceptual analogies between hereditary and infectious units, but it constituted a new instrument for moving and mapping bacterial genes. In 1955, Edward Lennox demonstrated that lysogenic phage P1 could transduce a variety of traits in *E. coli* and in *Shigella* (Lennox 1955; Brock 1990: 197). Co-transduction of alleles using phage P1 (for *E. coli*) or P22 (for *Salmonella*) provided a new test of linkage for bacterial genes, as only markers closely spaced on the chromosome would be packaged together into the same phage.

Transduction also fed into biochemical investigations of bacteria. Researchers found that for many of the major metabolic pathways, the bacterial genes controlling consecutive biosynthetic steps were linked. As a consequence, transduction could be used to analyze both the order of steps in biosynthesis and the mechanisms for their genetic control. Thus once again, advances in microbial genetics (and specifically, in this case, mapping techniques) contributed powerfully to biochemical investigations of metabolism, exemplified in studies of *Salmonella typhimurium* (Figure 1.7).

A few years later, Charles Yanofsky used P1 transduction in *E. coli* in conjunction with peptide mapping to compare mutations in the genetic locus for tryptophan synthetase with amino acid changes in the enzyme (Yanofsky *et al.* 1964).

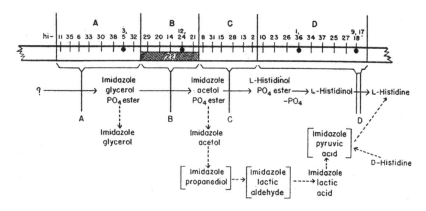

Figure 1.7 Linkage map of the histidine region of *Salmonella* chromosome. Solid arrows indicate the probable linkage relation of genetic blocks with the proposed biosynthetic pathway for L-histidine. The dotted arrows indicate probable secondary reactions. Reproduced with permission from Hartman, P.E. (1956) "Linked Loci in the Control of Consecutive Steps in the Primary Pathway of Histidine Synthesis in *Salmonella typhimurium*," in Carnegie Institution of Washington, Department of Genetics, *Genetic Studies with Bacteria*, Washington, DC: Carnegie Institution of Washington Publication 612: 41.

Not only did this provide a way to postulate nucleotide triplet–amino acid correlations for the genetic code, but the findings established definitively the molecular colinearity of gene and protein. In short, mapping had progressed to the molecular level. Fine-structure mapping was not an end to itself, however, but rather a tool to investigate the genetic code, metabolism, cellular regulation, and the dynamics of infection, to name but a few topics.

More broadly, the development of genetic mapping in bacteria and viruses subtly reshaped the conceptualization of genes, which no longer needed to reside on nuclear chromosomes to be mappable. Indeed, the lysogenic viruses, fertility factors, and episomes central to sexual exchange in bacteria became experimental tools for mapping bacterial genes—they substituted experimentally for the gametes of classical genetic techniques. Ultimately, the use of such genetic vectors (plasmids in Lederberg's broad sense) meant that all chromosomal genes were potentially mobile; the act of mapping itself required the transfer of genetic markers via cytoplasmic genetic elements. Seen from this perspective, the development of genetic engineering techniques for eukaryotes in the 1970s built on the experimental manipulation of prokaryotic genes that had been going on for more than a decade via experimental uses of transduction and conjugation.[55]

The anthropomorphism involved in treating bacterial conjugation as heterosexual reproduction has often been commented upon,[56] but comparatively little attention has been paid to how the study of genetic exchange in bacteria has destabilized conventional views of sex. The mapping of genes in bacteria relied on a shift in meaning from genetic exchange via *meiosis* to genetic exchange via agents of *infectious heredity*, and this revision in turn expanded the purview of genetic transmission in multicellular vertebrates—to include somatic mammalian cells as well as gametes. By the 1970s, biologists had begun investigating viruses as agents of quasi-sexual exchanges in mammalian cells. As Lederberg has remarked on these developments, "[w]e should be looking for sex where it had never been seen before" (Lederberg 1993: 43).

Concluding reflections

Jean-Paul Gaudillière has noted that American biology bears the "strong imprint of the Morgan group and its way of doing genetics," especially as reflected in collective patterns of work and the wide circulation of results, materials (especially mutant strains), and new methods (Gaudillière 2002: 406). This culture of scientific exchange has received ample attention in the historiography of the phage group,[57] but it was also a significant feature in the community of bacterial geneticists involved in mapping of microbes. Evelyn Witkin has recently discussed how the *Microbial Genetics Bulletin*, a mimeographed newsletter, came into circulation in 1950, following the example of the *Drosophila Information Service* and the *Phage Information Service* (Witkin 2002).[58] Many important findings, changes in terminology, and new methods were communicated in this bulletin, whose international circulation began with 74 individuals and grew to over one thousand by the 1970s. (The *Neurospora* research community spawned their own newsletter and established a stock center in the 1960s (Davis 2000: vii, 4–7).) Lederberg's accounts of his early experiments with *E. coli* emphasize the crucial importance of the Cold Spring Harbor meetings (especially that in 1946) in developing the social and intellectual network of bacterial geneticists, and illustrate the significance of material exchanges, particularly the nutritional auxotroph strains of *E. coli* K-12. After the announcement he and Tatum made of mating in *E. coli*, Lederberg found himself dealing with a "trickle (later a torrent) of requests for the cultures" (Lederberg 1987: 37).[59]

Yet, unlike Drosophilists, bacterial geneticists did not form a community based around a single model organism (despite Delbrück's limiting treaty for the phage group). Indeed, the field benefited from the multiplicity of model systems being investigated at any moment—*Neurospora crassa*, *Penicillium*, *E. coli*, *Salmonella typhirium*, and the smaller "model organisms" of the T-even phages, phages λ, P22, P1, and various other agents of infective heredity. Each experimental system provided a model inspiring research to find similar processes in other systems, attempts that often led to discoveries of new and different modes of genetic exchange.[60] Lederberg and Zinder turned to *Salmonella* to find patterns of genetic recombination like those in *E. coli*; their painstaking work resulted in observations of genetic exchange, but by a quite different process they termed transduction. In turn, transduction was also identified in *E. coli*, quite separate from conjugation by F factors, and this provided a new technological tool for "transforming" strains with new genetic markers. As bacterial geneticists multiplied examples of genetic exchange, the diversity of modes and mechanisms made a generic definition of bacterial sex elusive.

Even as fruit flies continued to serve as standard bearers for genetics,[61] bacterial geneticists provided strong competition for the Drosophilists. The availability of methods for fine-mapping of genetic structures and the ability to study even rare events by screening billions of bacterial colonies put microbes at the forefront of genetics research by the 1950s.[62] Whereas the early understanding of "gene" emanating from bacterial and viral genetics relied on defining a genetic unit in terms of mutation, rather than recombination, the power of recombinational analysis in bacterial genetics soon overtook that available in other organisms. As Joshua Lederberg put it in discussing the postwar period, "Bacteria and bacterial viruses quickly supplanted fruit flies as the test-bed for many of the subsequent developments of molecular genetics and the biotechnology that followed" (Lederberg 2000: 288).

Eventually, the tools and approaches generated through microbial genetics fed back into the study of some of the classic organisms, as the mapping of flies went molecular (Chapter 3 by Weber, this volume) and mice became genetic tools for, among other subfields, molecular immunology (Podolsky and Tauber 1997). By the 1980s, the trajectory from genetic mapping in microorganisms to the molecular genetics of eukaryotes was clearly associated with the molecularization of the gene. Yet as Marcel Weber has argued, the kind of fine-structure gene maps obtainable through bacteria and bacteriophage are not automatically molecular in nature (Weber 1998). Benzer, for instance, ascertained both the recombination map and complementation map of the rII region of T4 phage, and showed these two maps were in agreement, but did not provide independent evidence that these maps corresponded to actual DNA molecules. The molecular turn in mapping practices might be differentiated into two aspects. On the one hand, researchers after Benzer did use biophysical techniques to show that the fine-structure genetic maps represent physical features of DNA molecules. Weber (1998) points to Dale Kaiser's work with points of DNA breakage in phage λ and Charles Yanofsky's results with the tryptophan A gene of *E. coli*, mentioned earlier. But on the other hand, a second way in which the mapping practices of bacterial geneticists contributed to the molecularization of the gene was through the close association of gene mapping and biochemical research, with its pragmatic orientation to molecules as the agents of biological change. From the time of Beadle and Tatum's success in using genetic mutants to elucidate metabolism in *Neurospora*, a generation of microbiologists and biochemists used mapping techniques to advance biochemical knowledge, including the biochemistry of nucleic acids and gene regulation. I would suggest that the pervasive molecular view of genes in the postwar period reflected, in part, these extensive advances in biochemistry, especially those associated with microbial genetics. In the end, results substantiating the molecularization of the gene simply confirmed the workaday biochemical realism that accompanied the mapping of genes in viruses, bacteria, and fungi.

Acknowledgments

Research for this chapter was enabled by the author's NSF CAREER grant, SBE 98-75012. For useful responses and suggestions to an earlier draft of this chapter, I should like to thank Thomas Brock, Joshua Lederberg, Joseph November, Jean-Paul Gaudillière and Hans-Jörg Rheinberger. In addition, I am grateful to Evelyn Witkin for helpful conversations about bacterial genetics during the 1940s and 1950s. I thank Mrs Elizabeth Davis for allowing me to quote stanzas from Dr Bernard Davis's unpublished poem "On Sexing Bacteria."

Notes

1 See Dubos (1945), esp. 22–8 and "Addendum" (by C.F. Robinow on pp. 353–77). Robinow had produced the best microscopic evidence to date for "nucleoids" and chromosomes in bacteria, but opinion among bacteriologists was divided about whether these genetic structures existed in bacteria and what role they might play in heredity.

2 For example, see Blakeslee (1904).

3 On "nucleocentrism" and the debates over cytoplasmic inheritance, see Sapp (1987).

4 For an excellent overview of the development of bacteriology in France and Germany, see Mendelsohn (1996). My analysis throughout this chapter draws on various aspects of Thomas Brock's comprehensive and insightful account, *The Emergence of Bacterial Genetics* (Brock 1990).

5 On the development of Koch's notion of "pure culture," see Gradmann (2000).

6 As Christoph Gradmann puts it, for the period 1860–80, the notion "that these organisms could be sorted into distinct and constant species was—all in all—a minority opinion" (Gradmann 2000: 150).

7 As Mazumdar points out, Koch's demonstration that the anthrax bacillus had a life cycle including sporulation, and that there was a clear relationship between the bacterial spores and the disease, reinforced Cohn's view that bacteria were discrete species (Mazumdar 1995: 60–1).

8 For an overview of genetic studies of fungi, see Hayes (1965: ch. 4).

9 Thomas Brock quotes C. Flügge along these lines (Brock 1990: 30–1).

10 For a contemporary review of the literature, see Dienes (1946). I am indebted to Princeton graduate Joseph November, who recently surveyed this literature in a research paper (November, unpublished).

11 Amsterdamska pays particular attention to response to two defenders of the cyclogenic theory. In 1921, Felix Löhnis, a German bacteriologist who worked for the US Department of Agriculture, had reviewed the literature on bacterial life cycles, arguing strongly that most bacterial variation could be accounted for by the morphological and physiological changes associated with stages of development. Löhnis's radical cyclogenic interpretation involved a stage of "symplasm," a noncellular phase of bacterial life (which was presumably filterable) that could putatively reconstitute microbial cells. Related to Löhnis's intellectual agenda was his rejection of pure culture technique, standardized media, and many other new methods of bacteriological research, in favor of a natural historical approach within the laboratory. Other bacteriologists sympathetic to the theory, most notably Philip Hadley, attempted to employ biochemical and serological experimentation to advance a cyclogenic interpretation, but opponents mobilized these same techniques to argue against life cycle changes, especially in pointing out how morphological and physiological changes are not necessarily correlated. Thus, while the growing biochemical orientation of bacteriology did not itself settle debates over cyclogenic theory, the morphological orientation of bacteriology lost

ground as pure culture techniques and biochemical methods predominated. (Amsterdamska 1991: 197–8, 214–15). As November has pointed out (November, unpublished), Löhnis was one of the major early proponents of bacterial sexual organs—in the form of filter-passing gonidia.

12 Both Martinus Beijerinck in Holland and Sergie Winogradsky in Russia contributed to the development of synthetic culture media as part of their research efforts in agricultural bacteriology (Brock 1990: 32; Brock and Schlegel 1989).

13 On the rapid growth of research on bacterial nutrition in the 1930s, see Kohler (1985b: 62ff). Marjory Stephenson's investigations into the adaptive enzymes were especially prominent during the 1930s. Stephenson and her coworkers at Cambridge published several elegant experiments demonstrating that adaptation in their systems arose from the synthesis of completely new enzymes in the bacteria after the cells encountered the substrate, and not from selection and growth of mutant cells that harbored the enzyme (Stephenson 1938; Kohler 1985a). Dubos (1945: ch. 5) gives a valuable, nearly contemporary assessment of research on bacterial variability.

14 Many of the amino acids were commercially available by 1930; others could be straightforwardly isolated in the laboratory. The identification of vitamins such as thiamine (1936), riboflavin (1935), pyridoxine (1939), and biotin (1943) in turn enabled bacteriologists to screen cultures for dependence on these specific nutritional requirements (Brock 1990: 32). Knight obtained thiamine and two of its analogues from A.R. Todd and F. Bergel, who were studying the chemistry of this vitamin (B_1) (Knight 1971: 17). On vitamin research, see Kamminga (1998) and Kornberg (1989: ch. 1), "The vitamin hunters."

15 Francis Ryan cites twelve "instances of the adaptations of strains of bacteria and yeasts to dispense with complete growth-factor requirements," (Ryan 1946: 224). He also cites H.I. Kohn and J.S. Harris's generation of a methionine-requiring strain of *E. coli*, through repeated culturing in media containing methionine and sulfanilamide.

16 I offer an account of these debates elsewhere (Creager, unpublished).

17 The other major model organism for studying the physiological genetics of development at this time was the flour moth, *Ephestia*; see Rheinberger (2000).

18 Bernard O. Dodge of the Brooklyn Botanical Garden had analyzed the sexual life cycle of *Neurospora crassa*, and Carl Lindegren worked out the cytogenetics of the organism for his PhD dissertation (Kohler 1991: 119–20).

19 Lederberg (1960) provides the fullest biographical account of Tatum. It is worth emphasizing that he was well-trained in both biochemistry and bacteriology, having done his doctoral research in nutrition and microbial metabolism; see also Kohler (1991).

20 The separation of the eight spores generated by the initial cross required microscopic dissection of the spore sac. A nice description of the procedure is given by Kay (1989: 81).

21 Kohler (1991) argues that the wartime support of the *Neurospora* work reinforced the biochemical direction of its development, and also suggests that the relative failure of the genetic side of the program was due to technical problems with the cytogenetics. However, other accounts (e.g. R. Davis 2000: 3–7) do not represent the genetics in *Neurospora* as lagging. David Perkins at Stanford did much to consolidate and advance genetic research with *Neurospora* after the Second World War.

22 On criticisms of Beadle and Tatum's interpretations in the late 1940s and early 1950s, see Horowitz (1979).

23 That same year another group reported on x-ray-generated growth-factor deficient strains of *E. coli* as well (Roepke *et al.* 1944).

24 Bernard Davis introduced this terminology (to match Lederberg and Ryan's term prototroph) in the first *Microbial Genetics Bulletin*, which appeared in January 1950 (Witkin 2002). It was rapidly adopted in the bacterial genetics community.

25 Lederberg details their collaborations in his biographical memoir of Tatum (Lederberg 1960: 370ff).

26 The 1944 publication by Avery, Macleod, and McCarty on their identification of the pneumococcal transforming factor as DNA motivated Lederberg's interest in investigating sexual exchange in bacteria (Lederberg 1986).

27 Lederberg (1987: 34) notes that he and Ryan were encouraged in this venture by Beadle and Verna Coonradt's observation of nutritional symbiosis in *Neurospora* heterocaryons (Beadle and Coonradt 1944). Lederberg subsequently began to try to establish a system for examining exchange between strains of *Salmonella*, but the genetics of this bacterium were not well developed. A copy of his initial letter of inquiry to Tatum is reproduced in Lederberg (1987: 36).

28 As Lederberg put it, "[c]oincidental reversion at two or more loci is theoretically improbable, and experimentally undemonstrable" (Lederberg 1947: 509).

29 Lederberg and Ryan had introduced the term "prototroph" refer to such a strain "which has the nutritional requirements of the 'wild-type' from which it was derived irrespective of how it became prototrophic" (Ryan and Lederberg 1946: 172 fn 5).

30 As Lederberg recounts (1987), their paper was received with great interest, and the main source of skepticism was from Lwoff, who was concerned that cross-feeding might account for the prototrophs.

31 Indeed, Lederberg states that it was the excitement about the results of recombination in bacteriophage obtained by Hershey and independently by Delbrück that inspired Tatum and him to also announce their results for recombination in bacteria at the 1946 summer meeting (Lederberg 1993: 38).

32 Raphael Falk (1986) points to the importance of H.J. Muller in promoting a mutational and material view of the gene, as opposed to the more operational unit obtainable through recombination frequencies.

33 Salvador Luria had first identified a phage mutant (T2*h*) in 1944 (Luria 1945), and Hershey identified the second (*r*) in his Cold Spring Harbor paper (Hershey 1946).

34 As Lederberg points out, Dobzhansky offered this explanation for the Avery, Macleod, and McCarty experiments: "we are dealing with authentic cases of induction of specific mutations by specific treatments—a feat which geneticists have vainly tried to accomplish in higher organisms" (Dobzhansky 1947: 49, as quoted in Lederberg 1987: 30).

35 Based on the notion that there was a transient diploid zygote from mating, Lederberg and his colleagues proposed the following terminology: p-1 × p-1 (n)→F-1 (2n)→f-1 (n) (Lederberg *et al.* 1951: 416).

36 Lederberg later obtained DNase from Maclyn McCarty, and showed that it had no effect on the recovery of prototrophs, even though it had been shown to destroy the pneumococcal transforming factor (Lederberg *et al.* 1951: 414; Brock 1990: 84).

37 I take up Falk's analysis in a more extended way in discussing the attempts to use viruses as experimental tools for genetics in Creager (2002: ch. 6).

38 Davis devised an ingenious apparatus for this experiment, a U-shaped tube, with a fritted glass bacterial filter (i.e. impassable to cells) at the bottom separating the two arms. One could place two mating strains on the opposite sides of the filter to see if contact between the mating bacterial cells was required for mating. The same sort of apparatus was later used by Zinder and Lederberg to demonstrate that transduction, unlike mating, could be achieved without cell-to-cell contact (Brock 1990: 87–8).

39 As Brock (1990: 86) notes, the fact that *Paramecium* also provided one of the best examples of cytoplasmic inheritance (the *kappa* factor) meant that the simple analogy with protozoan sexual conjugation became enmeshed with debates over whether non-nuclear heredity might account for the development or transmission of certain traits (e.g., as associated with adaptation).

40 For a contemporary review of the literature, see Hutchinson and Stempen (1954). November (unpublished) provides a valuable historical appraisal.

41 Cavalli (who also published under Cavalli-Sforza) also isolated a strain of *E. coli* that mated at a high frequency; he called this strain Hfr (Cavalli-Sforza 1950).

42 By the time of the publication of his laboratory's 1951 Cold Spring Harbor paper, Lederberg had amassed the results from more extensive screening. The percentage of

fertile strains among those tested remained low; the laboratory reported that at least 20 of 650 strains from the Wisconsin Public Health Laboratory collection showed signs of recombination with a K-12 test strain (Lederberg *et al.* 1951: 437).

43 As the authors stated: "In a purely formalistic way, these data could be represented in terms of a 4-armed linkage group, Figure 1a, without supposing for a moment that this must represent the physical situation. This recalls the branched chromosome representation...of translocation heterozygotes in *Drosophila* before the cytogenetics of this situation was well understood" (Lederberg *et al.* 1951: 417).

44 Bernard D. Davis, "On Sexing Bacteria," unpublished, Joshua Lederberg Papers, "Profiles in Medicine," National Library of Medicine, Library of Congress, dated 1950. Available at <http://profiles.nlm.nih.gov/BB/> (accessed December 18, 2002).

45 Thomas Brock suggests that research on bacterial mating is conveniently divided into a pre-Hayes period and a post-Hayes period (Brock 1990: 87).

46 Hayes cites the work of André Lwoff, who had shown that the lytic phase of lysogenic infection could be induced by ultraviolet light, and of Jean Weigle and Max Delbrück, who extended this observation to show that *E. coli* K-12 was lysogenized by inducible phage λ (Hayes 1952b).

47 On the historical and cultural association of masculinity with agency and femininity with passivity, see Keller (1985). Accounts of gender in the history of bacterial mating are given by Sapp (1990: ch. 3), Spanier (1995: 56–9), and Bivins (2000). Bivins points to the long history of attributing more agency to sperm than to eggs in reproduction (for one insightful analysis, see (Martin 1991)). At the same time, for the sake of formal genetics—which seems to have been the most significant source of inspiration for bacterial geneticists—the contributions from father and mother are equivalent.

48 This notion (*plasmatische Gene*) was already being used widely in Germany, as introduced by Hans Winkler (1924). An excellent account of the German research into cytoplasmic investigation, including Winkler's contributions, is offered by Jonathan Harwood (1984).

49 The interest among microbiologists in the gene–enzyme relation is registered by Guido Pontecorvo in his review of 1945 conference papers on the theme "Gene Action in Microorganisms" (Pontecorvo 1946).

50 He urged that the various genetic units outside the nucleus be considered along a spectrum of "infective heredity," with "deleterious parasitic viruses at one extreme, and integrated cytoplasmic genes like plastids, at the other. Within this interval, we find a host of transition forms; kappa, lysogenic bacteriophages, genoids, tumor-viroids, male-sterility factors, Ephrussi's yeast granules, etc.... The objection has been voiced that this viewpoint is an attempt to relegate plasmagenes to pathology. I rather think that it may broaden our genetic point of view if we consider the likenesses as well as the dissimilarities between pathogenic viruses and plasmagenes." He contended that lysogenic bacteriophage provided "the best material" for investigating the genetic character of such cytoplasmic genes, "especially as infection with such a virus is formally indistinguishable from events such as pneumococcus transformations" (Lederberg 1951b: 286, 1998).

51 Emma Klieneberger-Nobel and Louis Dienes had described L forms and debated their nature since the 1930s. Lederberg kept up with this literature on the morphology of bacteria and bacterial life cycles; in their 1952 paper Zinder and Lederberg cite Dienes and Weinberger (1951), Klieneberger-Nobel (1951), and Tulasne (1951).

52 On the naming, see Lederberg *et al.* (1951: 436). As Brock notes, "cytogenes had traditionally been designated by Greek letters" (Brock 1990: 179).

53 Jacob and Wollman (1961) give an extended description of the findings and full bibliography; the developments are also covered by Brock (1990).

54 Carrying the analogy of bacterial genetic exchange with human heterosexuality further, the French team referred to this technique as "coitus interruptus" (Lederberg 1993: 39).

55 One might suggest that molecular geneticists studying tumor viruses used transformation similarly as a way to manipulate and move genes in eukaryotes. See Morange (1998: ch. 16).

56 For example, Sapp (1990); Spanier (1995); Bivins (2000).

57 Morange (1998) provides a useful synopsis of the phage group and its historiography.
58 Several of the findings discussed in this paper—such as Esther Lederberg's identifica-
    tion of *E. coli* K-12 as lysogenic (E. Lederberg 1951) and Bernard Davis's observation
    that cell-to-cell contact was required for mating in *E. coli* (B. Davis 1950)—were first
    reported in this bulletin.
59 Gaudillière (2002: 407) analyzes the frustrations Lederberg faced in trying to establish
    free circulation of information and strains with the Pastorians (especially Monod).
60 I have argued elsewhere for the significance of particular research systems as exemplars
    in this sense to research in molecular biology (Creager 2002: ch. 8).
61 I owe this notion of model organisms as "standard-bearers" to Karen Rader (2004).
62 For one geneticist's perspective on the importance of findings with microbes to 1950s
    genetics at large, see Pontecorvo (1958).

Bibliography

Amsterdamska, O. (1987) "Medical and Biological Constraints: Early Research on
    Variation in Bacteriology," *Social Studies of Science*, 17: 657–87.
——— (1991) "Stabilizing Instability: The Controversy over Cyclogenic Theories of Bacterial
    Variation during the Interwar Period," *Journal of the History of Biology*, 24: 191–222.
Anderson, E.H. (1944) "Incidence of Metabolic Changes among Virus-Resistant Mutants
    of a Bacterial Strain," *Proceedings of the National Academy of Sciences USA*, 30: 397–403.
—— (1946) "Growth Requirements of Virus-Resistant Mutants of *Escherichia coli* Strain
    'B'," *Proceedings of the National Academy of Sciences USA*, 32: 120–8.
Avery, O.T., MacLeod, C.M., and McCarty, M. (1944) "Studies on the Chemical Nature
    of the Substance Inducing Transformation of Pneumococcal Type: Induction of
    Transformation by a Deoxyribonucleic Acid Fraction Isolated from Pneumococcus
    Type III," *Journal of Experimental Medicine*, 79: 137–58.
Beadle, G.W. (1945a) "Biochemical Genetics," *Chemical Reviews*, 37: 15–96.
—— (1945b) "Genetics and Metabolism in *Neurospora*," *Physiological Reviews*, 25: 643–63.
—— (1974) "Recollections," *Annual Review of Biochemistry*, 43: 1–13.
Beadle, G.W. and Coonradt, V.L. (1944) "Heterocaryosis in *Neurospora crassa*," *Genetics*, 29:
    291–308.
Beadle, G.W. and Ephrussi, B. (1936) "The Differentiation of Eye Pigments in *Drosophila* as
    Studied by Transplantation," *Genetics*, 21: 225–47.
—— (1937) "Development of Eye Colors in *Drosophila*: Diffusible Substances and their
    Interrelations," *Genetics*, 22: 76–86.
Beadle, G.W. and Tatum, E.L. (1941) "Genetic Control of Biochemical Reactions in
    Neurospora," *Proceedings of the National Academy of Sciences USA*, 27: 499–506.
Bivins, R. (2000) "Sex Cells: Gender and the Language of Bacterial Genetics," *Journal of
    the History of Biology*, 33: 113–39.
Blakeslee, A.F. (1904) "Sexual Reproduction in the *Mucorineae*," *Proceedings of the American
    Academy of Arts and Sciences*, 40: 205–319.
Brock, T.D. (1990) *The Emergence of Bacterial Genetics*, Cold Spring Harbor, NY: Cold Spring
    Harbor Laboratory Press.
Brock, T.D. and Schlegel, H.G (1989) "Introduction," in H.G. Schlegel and B. Bowien
    (eds), *Autotrophic Bacteria*, Madison, WI: Science Tech. Publishers, 1–15.
Cairns, J. (1963) "The Chromosome of *Escherichia coli*," *Cold Spring Harbor Symposia on
    Quantitative Biology*, 28: 43–6.
Cavalli-Sforza, L.L. (1950) "La sessualità nei batteri," *Bollettino Istituto Sieroterapico Milanese*,
    29: 281–9.

Cavalli, L.L. and Heslot, H. (1949) "Recombination in Bacteria: Outcrossing *Escherichia coli* K 12," *Nature*, 164: 1057–8.

Cavalli, L.L., Lederberg, J., and Lederberg, E.M. (1953) "An Infective Factor Controlling Sex Compatibility in *Bacterium coli*," *Journal of General Microbiology*, 8: 89–103.

Creager, A.N.H. (2002) *The Life of a Virus: Tobacco Mosaic Virus as an Experimental Model, 1930–1965*, Chicago, IL: University of Chicago Press.

—— "Adaptation or Selection? Old Issues and New Stakes in the Postwar Debates about Microbial Drug Resistance," unpublished paper presented at the Department of the History of Science, Medicine, and Technology, Johns Hopkins University on October 17, 2002.

Darlington, C.D. (1944) "Heredity, Development and Infection," *Nature*, 154: 164–9.

Davis, B.D. (1948) "Isolation of Biochemically Deficient Mutants of Bacteria by Penicillin," *Journal of the American Chemical Society*, 70: 4267.

—— (1950) "Nonfiltrability of the Agents of Genetic Recombination in *Escherichia coli*," *Journal of Bacteriology*, 60: 507–8.

Davis, R.H. (2000) *Neurospora: Contributions of a Model Organism*, Oxford: Oxford University Press.

Delbrück, M. and Bailey, W.T. (1946) "Induced Mutations in Bacterial Viruses," *Cold Spring Harbor Symposia on Quantitative Biology*, 11: 33–7.

Dienert, F. (1900) "Sur la fermentation du galactose et sur l'accutunance des levures à ce sucre," *Annales de l'Institut Pasteur, Paris*, 14: 139–89.

Dienes, L. (1946) "Complex Reproductive Processes in Bacteria," *Cold Spring Harbor Symposia on Quantitative Biology*, 11: 51–9.

Dienes, L. and Weinberger, H. J. (1951) "The L Forms of Bacteria," *Bacteriological Reviews*, 15: 245–88.

Dobzhansky, T. (1947) *Genetics and the Origin of Species*, 2nd edn, New York: Columbia University Press.

Doermann, A.H., Chase, M., and Stahl, F.W. (1955) "Genetic Recombination and Replication in Bacteriophage," *Journal of Cellular and Comparative Pathology*, 45, Supplement 2: 51–74.

Doudoroff, M. (1940) "Experiments on the Adaptation of *Escherichia coli* to Sodium Chloride," *Journal of General Physiology*, 23: 585–611.

Dubos, R. (1945) *The Bacterial Cell in its Relation to Problems of Virulence, Immunity and Chemotherapy*, Cambridge, MA: Harvard University Press.

Ephrussi, B. and Beadle, G.W. (1937) "Development of Eye Colors in *Drosophila*: Transplantation Experiments on the Interaction of Vermillion with other Eye Colors," *Genetics*, 22: 65–75.

Falk, R. (1986) "What is a Gene?" *Studies in History and Philosophy of Science*, 17: 133–73.

Fried, P.J. and Lederberg, J. (1952) "Linkage in *E. coli* K-12 (abstract)," *Genetics*, 37: 582.

Gaudillière, J.-P. (1992) "J. Monod, S. Spiegelman et l'adaptation enzymatique. Programmes de recherche, cultures locales et traditions disciplinaires," *History and Philosophy of the Life Sciences*, 14: 23–71.

—— (2002) "Paris–New York Roundtrip: Transatlantic Crossings and the Reconstruction of the Biological Sciences in Post-War France," *Studies in History and Philosophy of Biological and Biomedical Sciences*, 33C: 389–417.

Gradmann, C. (2000) "Isolation, Contamination, and Pure Culture: Monomorphism and Polymorphism of Pathogenic Microorganisms as Research Problem, 1860–1880," *Perspectives on Science*, 9: 147–72.

Gray, C.H. and Tatum, E.L. (1944) "X-Ray Induced Growth Factor Requirements in Bacteria," *Proceedings of the National Academy of Sciences USA*, 30: 404–10.

Hartman, P.E. (1956) "Linked Loci in the Control of Consecutive Steps in the Primary Pathway of Histidine Synthesis in *Salmonella typhimurium*," in Carnegie Institution of Washington, Department of Genetics, *Genetic Studies with Bacteria*, Washington, DC: Carnegie Institution of Washington, 35–62.

Harwood, J. (1984) "The Reception of Morgan's Chromosome Theory in Germany: Inter-War Debate over Cytoplasmic Inheritance," *Medizinhistorisches Journal*, 19: 3–32.

Hayes, W. (1952a) "Recombination in *Bact. coli* K-12: Unidirectional Transfer of Genetic Material," *Nature*, 169: 118–19.

—— (1952b) "Genetic Recombination in *Bact. coli* K-12: Analysis of the Stimulating Effect of Ultra-Violet Light," *Nature*, 169: 1017–18.

—— (1953) "Observations on a Transmissible Agent Determining Sexual Differentiation in *Bacterium coli*," *Journal of General Microbiology*, 8: 72–88.

—— (1965) *The Genetics of Bacteria and Their Viruses: Studies in Basic Genetics and Molecular Biology*, New York: John Wiley & Sons.

Hershey, A.D. (1946) "Spontaneous Mutations in Bacterial Viruses," *Cold Spring Harbor Symposia on Quantitative Biology*, 11: 67–77.

Hershey, A.D. and Rotman, R. (1948) "Linkage Among Genes Controlling Inhibition of Lysis in a Bacterial Virus," *Proceedings of the National Academy of Sciences USA*, 34: 89–96.

—— (1949) "Genetic Recombination between Host-Range and Plaque-Type Mutants of Bacteriophage in Single Bacterial Cells," *Genetics*, 34: 44–71.

Holmes, F.L. (2000) "Seymour Benzer and the Definition of the Gene," in P. Beurton, R. Falk, and H.-J. Rheinberger (eds) *The Concept of the Gene in Development and Evolution: Historical and Epistemological Perspective*, Cambridge: Cambridge University Press, 115–55.

—— (2001) *Meselson, Stahl, and the Replication of DNA: A History of 'The Most Beautiful Experiment in Biology,'* New Haven, CT: Yale University Press.

Horowitz, N.H. (1979) "Genetics and the Synthesis of Proteins," in P.R. Srinivasan, J.S. Fruton, and J.T. Edsall (eds) *The Origins of Modern Biochemistry: A Retrospect on Proteins*, New York: New York Academy of Sciences, 253–62.

Hutchinson, W.G. and Stempen, H. (1954) "Sex in Bacteria: Evidence from Morphology," in D.H. Wenrich, I.F. Lewis, and J.R. Raper (eds) *Sex in Microorganisms*, Washington, DC: American Association for the Advancement of Science, 29–41.

Jacob, F. and Wollman, E.L. (1961) *Sexuality and the Genetics of Bacteria*, New York: Academic Press.

Kamminga, H. (1998) "Vitamins and the Dynamics of Molecularization: Biochemistry, Policy and Industry in Britain, 1914–1939," in S. de Chadarevian and H. Kamminga (eds) *Molecularizing Biology and Medicine: New Practices and Alliances, 1910s–1970s*, Amsterdam: Harwood Academic Publishers, 83–105.

Karström, H. (1937) "Enzymatische Adaptation bei Mikroorganismen," *Ergebnisse der Enzymforschung*, 7: 350–76.

Kay, L.E. (1989) "Selling Pure Science in Wartime: The Biochemical Genetics of G.W. Beadle," *Journal of the History of Biology*, 22: 73–101.

Keller, E.F. (1985) *Reflections on Gender and Science*, New Haven: Yale University Press.

Klieneberger-Nobel, E. (1951) "Filterable Forms of Bacteria," *Bacteriological Reviews* 15: 77–130.

Knight, B.C.J.G. (1971) "On the Origins of 'Growth Factors'," in J. Monod and E. Borek (eds) *Of Microbes and Life*, New York: Columbia University Press, 16–18.

Kohler, R.E. (1985a) "Innovation in Normal Science: Bacterial Physiology," *Isis*, 76: 162–81.

—— (1985b) "Bacterial Physiology: The Medical Context," *Bulletin of the History of Medicine*, 59: 54–74.

—— (1991) "Systems of Production: *Drosophila, Neurospora,* and Biochemical Genetics," *Historical Studies in the Physical and Biological Sciences,* 22: 87–130.

—— (1994) *Lords of the Fly:* Drosophila *Genetics and the Experimental Life,* Chicago, IL: University of Chicago Press, 1994.

Kornberg, A. (1989) *For the Love of Enzymes: The Odyssey of a Biochemist,* Cambridge, MA: Harvard University Press.

Krementsov, N. (1996) "A 'Second Front' in Soviet Genetics: The International Dimension of the Lysenko Controversy, 1944–1947," *Journal of the History of Biology,* 29: 229–50.

Lederberg, E.M. (1951) "Lysogenicity in *E. coli* K-12 (Abstract)," *Genetics,* 36: 560.

Lederberg, E.M. and Lederberg, J. (1953) "Genetic Studies of Lysogenicity in *Escherichia coli,*" *Genetics,* 38: 51–64.

Lederberg, J. (1947) "Gene Recombination and Linked Segregations in *Escherichia coli,*" *Genetics,* 32: 505–25.

—— (1951a) "Prevalence of *Escherichia coli* Strains Exhibiting Genetic Recombination," *Science,* 114: 68–9.

—— (1951b) "Genetic Studies with Bacteria," in L.C. Dunn (ed.) *Genetics in the 20th Century: Essays on the Progress of Genetics During its First Fifty Years,* New York: Macmillan.

—— (1952) "Cell Genetics and Hereditary Symbiosis," *Physiological Reviews,* 32: 403–30.

—— (1956) "Genetic Transduction," *American Scientist,* 44: 264–80.

—— (1960) "Edward Lawrie Tatum," *Biographical Memoirs of the National Academy of Sciences,* 59: 357–86.

—— (1986) "Forty Years of Genetic Recombination in Bacteria," *Nature,* 324: 627–8.

—— (1987) "Genetic Recombination in Bacteria: A Discovery Account," *Annual Review of Genetics,* 21: 23–46.

—— (1993) "Genetic Maps—Fruit Flies, People, Bacteria, and Molecules: A Tribute to Morgan and Sturtevant," in R.B. Barlow, Jr, J.E. Dowling, and G. Weissman (eds) *The Biological Century: Friday Evening Talks at the Marine Biological Laboratory,* Woods Hole, MA: The Marine Biological Laboratory, 26–49.

—— (1996) "Genetic Recombination in *Escherichia coli*: Disputation at Cold Spring Harbor, 1946–1996," *Genetics,* 144: 439–43.

—— (1998) "Plasmid (1952–1997)," *Plasmid,* 39: 1–9.

—— (2000) "Infectious History," *Science,* 288: 287–93.

Lederberg, J. and Tatum, E.L. (1946a) "Detection of Biochemical Mutants of Microorganisms," *Journal of Biological Chemistry,* 165: 381–2.

—— (1946b) "Novel Genotypes in Mixed Cultures of Biochemical Mutants of Bacteria," *Cold Spring Harbor Symposia on Quantitative Biology,* 11: 113–14.

—— (1946c) "Gene Recombination in *Escherichia coli,*" *Nature,* 158: 558.

—— (1953) "Sex in Bacteria: Genetic Studies, 1945–1952," *Science,* 118: 169–75; reprinted in D.H. Wenrich, I.F. Lewis, and J.R. Raper (eds) (1954) *Sex in Microorganisms,* Washington, DC: American Association for the Advancement of Science, 12–28.

Lederberg, J. and Zinder, N. (1948) "Concentration of Biochemical Mutants of Bacteria with Penicillin," *Journal of the American Chemical Society,* 70: 4267–8.

Lederberg, J., Lederberg, E.M., Zinder, N.D., and Lively, E.R. (1951) "Recombination Analysis of Bacterial Heredity," *Cold Spring Harbor Symposia on Quantitative Biology,* 16: 413–43.

Lederberg, J., Cavalli, L.L., and Lederberg, E.M. (1952) "Sex Compatibility in *Escherichia coli,*" *Genetics,* 37: 720–30.

Lennox, E.S. (1955) "Transduction of Linked Genetic Characters of the Host by Bacteriophage P1," *Virology,* 1: 190–206.

Luria, S.E. (1945) "Mutations of Bacterial Viruses Affecting their Host Range," *Genetics,* 30: 84–99.

Luria, S.E. and Delbrück, M. (1943) "Mutations of Bacteria from Virus Sensitivity to Virus Resistance," *Genetics*, 28: 491–511.

Lwoff, A. (1946) "Some Problems Connected with Spontaneous Biochemical Mutants in Bacteria," *Cold Spring Harbor Symposia on Quantitative Biology*, 11: 139–53.

—— (1953) "Lysogeny," *Bacteriological Reviews*, 17: 269–337.

Martin, E. (1991) "The Egg and the Sperm: How Science has Created a Romance Based on Stereotypical Male–Female Roles," *Signs*, 16: 485–501.

Mazumdar, P.M.H. (1995) *Species and Specificity: An Interpretation of the History of Immunology*, Cambridge: Cambridge University Press.

Medawar, P.B. (1947) "Cellular Inheritance and Transformation," *Biological Reviews*, 22: 360–89.

Mendelsohn, J.A. (1996) *Cultures of Bacteriology: Formation and Transformation of a Science in France and Germany, 1870–1914*, PhD Dissertation, Princeton University.

Morange, M. (1998) *A History of Molecular Biology*, trans. M. Cobb, Cambridge, MA: Harvard University Press.

Muller, H.J. (1951) "The Development of the Gene Theory," in L.C. Dunn (ed.) *Genetics in the 20th Century: Essays on the Progress of Genetics During its First Fifty Years*, New York: Macmillan, 77–99.

Newcombe, H.B. and Nyholm, M.H. (1950) "Anomalous Segregation in Crosses of *Escherichia coli*," *American Naturalist*, 84: 457–65.

November, J. "Seeing Genetic Bacteria: The Search for Visible Bacterial Fusion," unpublished research paper, spring 2002.

Podolsky, S.H. and Tauber, A.I. (1997) *The Generation of Diversity: Clonal Selection Theory and the Rise of Molecular Immunology*, Cambridge, MA: Harvard University Press.

Pontecorvo, G. (1946) "Microbiology, Biochemistry, and the Genetics of Microorganisms," *Nature*, 157: 95–6.

—— (1958) *Trends in Genetic Analysis*, New York: Columbia University Press.

Rader, K.A. (2004) *Making Mice: Standardizing Animals for American Biomedical Research, 1900–1955*, Princeton, NJ: Princeton University Press.

Rheinberger, H.-J. (2000) "*Ephestia*: The Experimental Design of Alfred Kühn's Physiological Developmental Genetics," *Journal of the History of Biology*, 33: 535–76.

Roepke, R.R., Libby, R.L., and Small, M.H. (1944) "Mutation or Variation of *Escherichia coli* with Respect to Growth Requirements," *Journal of Bacteriology*, 48: 401–12.

Ryan, F.J. (1946) "Back-Mutation and Adaptation of Nutritional Mutants," *Cold Spring Harbor Symposia on Quantitative Biology*, 11: 215–26.

Ryan, F.J. and Lederberg, J. (1946) "Reverse-Mutation and Adaptation in Leucineless *Neurospora*," *Proceedings of the National Academy of Sciences USA*, 32: 163–73.

Sapp, J. (1987) *Beyond the Gene: Cytoplasmic Inheritance and the Struggle for Authority in Genetics*, New York: Oxford University Press.

—— (1990) *Where the Truth Lies: Franz Moewus and the Origins of Molecular Biology*, Cambridge: Cambridge University Press.

Singleton, R., Jr. (2000) "From Bacteriology to Biochemistry: Albert Jan Kluyver and Chester Werkman at Iowa State," *Journal of the History of Biology*, 33: 141–80.

Snell, E.E. (1951) "Bacterial Nutrition—Chemical Factors," in C.H. Werkman and P.W. Wilson (eds) *Bacterial Physiology*, New York: Academic Press, 214–55.

Spanier, B.B. (1995) *Im/Partial Science: Gender Ideology in Molecular Biology*, Bloomington, IN: Indiana University Press.

Spath, S.B. (1999) *C. B. van Niel and the Culture of Microbiology*, PhD Dissertation, University of California, Berkeley.

Spiegelman, S. and Kamen, M.D. (1946) "Genes and Nucleoproteins in the Synthesis of Enzymes," *Science*, 104: 581–4.

Stephenson, M. (1938) "The Economy of the Bacterial Cell," in J. Needham and D.E. Green (eds) *Perspectives in Biochemistry*, Cambridge: Cambridge University Press, 91–8.

Tatum, E.L. and Beadle, G.W. (1942) "Genetic Control of Biochemical Reactions in *Neurospora*: An 'Aminobenzoicless' Mutant," *Proceedings of the National Academy of Sciences USA*, 28: 234–43.

Tulasne, R. (1951) "Les formes L des bactéria," *Revue d'immunologie et de thérapie antimicrobienne*, 15: 223–51.

Weber, M. (1998) "Representing Genes: Classical Mapping Techniques and the Growth of Genetical Knowledge," *Studies in History and Philosophy of Biological and Biomedical Sciences*, 29C: 295–315.

Winge, O. and Laustsen, O. (1937) "On Two Types of Spore Germination, and on Genetic Segregations in *Saccharomyces*, Demonstrated through Single-Spore Cultures," *Comptes-rendus des travaux du Laboratoire Carlsberg, Série physiologie* 22: 99–117.

Winkler, H. (1924) "Über die Rolle von Kern und Protoplasma bei der Vererbung," *Zeitschrift für induktive Abstammungs- und Vererbungslehre*, 33: 238–53.

Witkin, E. (2002) "Chances and Choices: Cold Spring Harbor, 1944–1955," *Annual Review of Microbiology*, 56: 1–15.

Wollman, E.L., Jacob, F., and Hayes, W. (1956) "Conjugation and Genetic Recombination in *Escherichia coli* K-12," *Cold Spring Harbor Symposium of Quantitative Biology*, 21: 141–62.

Wright, S. (1941) "The Physiology of the Gene," *Physiological Reviews*, 21: 487–527.

—— (1945) "Genes as Physiological Agents," *American Naturalist*, 79: 289–303.

Yanofsky, C., Carlton, B.C., Guest, J.R., Helsinki, D.R., and Henning, U. (1964) "On the Colinearity of Gene Structure and Protein Structure," *Proceedings of the National Academy of Sciences USA*, 51: 266–72.

Zinder, N.D. (1955) "Bacterial Transduction," *Journal of Cellular and Comparative Physiology* 45, Supplement 2: 23–49.

—— (2002) "The Discovery of Transduction," in *Great Experiments: A Series of Essays by Prominent Biologists*, published online by Ergito. Available at <http://www.ergito.com/lookup.jsp?expt = zinder> (accessed December 12, 2002).

Zinder, N.D. and Lederberg, J. (1952) "Genetic Exchange in *Salmonella*," *Journal of Bacteriology*, 64: 679–99.

2 Seymour Benzer and the convergence of molecular biology with classical genetics

Frederic L. Holmes

I

Between 1954 and 1961 Seymour Benzer mapped the fine structure of the rII region of the genome of the bacteriophage T4. This work achieved a historical significance far beyond the small bit of genetic material on which it was carried out. In the early years of the "double helix," Benzer's evidence that the genetic map is linear down to dimensions approaching those of nucleotides provided powerful and timely support for the hypothesis that information embedded in the order of the DNA base pairs codes for the order of amino acids in proteins. His evidence that the units of recombination, mutation, and function associated with the gene had different dimensions led him to propose three new terms to replace the "classical gene" with more rigorously defined concepts. Yet the broader similarity between his fine structure maps and those that had earlier been constructed for the location of genes on the chromosomes of organisms such as the fruit fly was striking enough so that Benzer has been seen as the person who, more than anyone else, provided the "bridge" that enabled classical genetics to "adapt to the molecular age."

Benzer has saved an astonishingly full record of his investigative pathway during these years. There is a complete series of research notebooks documenting all of the experiments he performed, a rich correspondence with colleagues, notes on meetings he attended, outlines of his own lectures and seminars, and reports on the various trips he took. From this material it will be possible to reconstruct his entire professional life during the time of his quest in exceptionally comprehensive detail. Here, however, I want to focus on a more limited question. To what extent did Benzer foresee, or set out to attain, such far-reaching effects when he took up his investigation of the rII region? Horace Judson has framed the situation in a way that implies considerable prior vision of the road ahead. "The problem Benzer proceeded to solve was a version of the ultimate classical problem: to map the gene completely and so to drive formal genetic analysis ... down to the level of the chemical gene" (Judson 1979: 271).

That the "problem" that Benzer eventually solved was the same one that he intended from the beginning to solve may, however, be an artifact of hindsight. Hans-Jörg Rheinberger has pointed out how difficult it is to avoid reconstructions that merge outcome with intent:

> Once a surprising result has emerged, has proved to be more than of an ephemeral character, and has been sufficiently stabilized, it becomes more and more difficult, even for the participants, to avoid the illusion that it is the inevitable product of logical inquiry.
>
> (Rheinberger 1997: 74)

How matters appear to participants before a result has emerged is vividly captured in a quotation Rheinberger takes from the evocative writing of François Jacob:

> What we can guess today will not be realized. Change is bound to occur anyway, but the future will be different from what we believe. This is especially true in science. The search for knowledge is an endless process and one can never tell how it is going to turn out…If what is to be found is really new, then it is by definition unknown in advance. There is no way of telling where a particular line of research will lead.
>
> (Rheinberger 1997: 182)

Rheinberger attributes the unpredictability of experimental science to the fact that the experimental system that a scientist devises to solve one problem takes on a life of its own, carrying its inventor toward previously unanticipated destinations. That this does happen Rheinberger amply documents in the case of the system that Paul Zamecnik and his associates designed to study the mechanism of protein synthesis, and which eventually led them to the characterization of an object—transfer RNA—whose existence they did not at the beginning suspect. This is, however, only a special form of the broader unpredictability that Jacob meant to attribute, not only to the nature of science, but to the human condition as a whole.

My doubt that he started out with the goal of carrying classical genetic mapping down to the level of the chemical gene clearly in mind was aroused when Benzer mentioned to me that he never had taken a course in genetics, and knew little about the classical mapping tradition when he began his research on the rII region. Several years before that, someone had told him about the work of the Morgan school, and he had thought it exciting, but had not followed up by reading their publications or other detailed discussions on the topic. Today I want to sketch a few of the first steps that led Benzer into his investigation of bacteriophage genetics—steps which suggest that, as in most explorations of the unknown, where he was headed only became fully clear to him along the way.

II

Benzer belongs to that prominent cohort among the "phage group" who were trained as physicists and inspired by Erwin Schrödinger's *What is Life* to consider applying their knowledge of physics, and the ways in which physicists think, to the solution of biological problems. Chromosomes, according to Schrödinger, "contain in some kind of code-script the entire pattern of the individual's future development and of its functioning in the mature state." Schrödinger proposed that the gene is a very "complicated molecule in which every atom, and every group of atoms, plays an individual role, not entirely equivalent to that of many others (as is the case in a periodic structure)." He particularly emphasized the way in which x-rays produce mutations. It was "fairly obvious," he wrote, that the "single event, causing a mutation, is just an ionization (or similar process) occurring within some 'critical' volume of the germ cell." From an observed mutation rate, which he picked out only for "illustration," he estimated that the "critical volume, the 'target' which has to be 'hit' by an ionization for that mutation to occur, is only... one fifty-millionth of a c.cm" (Schrödinger 1946: 20, 44, 61, 65–6).

When *What is Life* appeared in 1946, Benzer was a doctoral candidate in physics at Purdue University, participating in research on the use of germanium crystals as semi-conductors. A fellow student lent him a copy that same year. According to his later recollection, Benzer was thrilled particularly about Schrödinger's discussion of the remarkable permanence of gene molecules, and wondered if the properties of his germanium crystals might have something to do with the "aperiodic crystals" that Schrödinger proposed to explain the properties of genes. It is possible also that the discussion of x-rays, mutations, and "targets" especially appealed to Benzer as a physicist. Whatever may have been the particular ideas of Schrödinger that struck him most immediately was of less lasting significance, however, than the general curiosity it aroused in him about how physics can be applied to fundamental biological problems. Soon afterward, at a dinner party to which he had been invited, Benzer asked his host, Salvador Luria, about Max Delbrück, to whose "molecular model" of the gene Schrödinger had prominently referred. Luria, who had been collaborating with Delbrück for several years in research on bacteriophage, urged Benzer to attend the summer course at Cold Spring Harbor that Delbrück had established (Weiner 1999: 42–5).

After receiving his PhD in 1947, Benzer was appointed an instructor in physics. At the end of the academic year 1947–48, however, he followed the advice of Luria and drove with his wife Dotty, and their baby daughter Barbie, to Cold Spring Harbor, where the course began on June 28. Delbrück was not there that year, and the course lectures were given by Mark Adams. The thirty young scientists taking the course were divided into groups in which they carried out the fundamental experimental procedures of phage research: after learning to distinguish the seven types of phage "T1... T7" preferred by the phage group, they infected *E. coli* bacteria with each of the types, and followed the phases of the phage reproductive cycle—adsorption, penetration, multiplication, and lysis—by means of the changes in turbidity of the culture, and through microscopic observations. They learned to recognize the plaque sizes characterizing the different phage and mutant types, and familiarized themselves with the notations phage biologists used to designate them: "r", Benzer wrote in his notebook, "stands for rapid lysis (mutant)." "r$^+$ stands for not rapid lysis (wild type)."[1]

Further into the course they performed "mixed infection" experiments with two types of phage, and multiple infections by increasing the ratio of phage to bacteria. They inactivated phage with antibody serum, and irradiated infected bacteria with ultraviolet light. They learned that Raymond Laterjet had "irradiated bacteria and found that you can have a lethal mutation where the bacteria grow but cannot divide. These are susceptible to phage." They concentrated their attention on the staple fixture of phage research, the "single step growth curve," measuring latent periods and rates of lysis, and counting plaques to determine the distributions of burst sizes. They used the Poisson distribution equation to calculate the average multiplicity of infections.[2]

Adams's lectures were straightforward and not particularly inspiring to Benzer, but the total immersion in phage experimentation immediately captivated him. The experiments were simple, quantitative, and decisive. They could be finished in one day, and then the group was ready to move on to the next one. The interaction with other students and guest lecturers was stimulating. There was a shared feeling that they were being introduced to something new and beautiful. By the time the three-week course ended with the traditional graduation ceremony, Benzer had already decided to become a biologist. What he had learned there provided the foundation for much of what he did in the next several years (Weiner 1999: 45; Benzer, interview by the author, January 3, 2001, recorded, tape 1A).[3]

Obtaining a postdoctoral fellowship from the Atomic Energy Commission, and a leave of absence from Purdue, Benzer journeyed to the Oak Ridge National Laboratory in September, 1948 to begin his research career in bacteriophage biology. There he was attracted to the recently discovered phenomenon of "photoreactivation." Phage that had been inactivated by ultraviolet light after their adsorption on bacteria could be reactivated by exposing the bacteria to visible light. This observation appealed to him because it was "related to physics" (Benzer/Holmes, interview, January 3, 2001). Soon, however, he turned to another question involving the ultraviolet radiation of bacteriophage during their intracellular growth stage, one directly inspired by a study published in 1948 by Luria and Laterjet. The so-called "dark" stage between the time at which the phage particle penetrated a bacterium, and the release of its progeny through the lysis of the bacterium, remained mysterious. As Luria and Laterjet put it, "The process of intracellular phage growth—in particular, of the kinetics of phage production—has so far escaped every attempt at clarification made either by breaking down infected bacteria or by electron microscopy." Their approach, to irradiate the infected bacteria at several time intervals during this phase, with three different doses of ultraviolet light, in order to determine the changes in the resistance of the "infective centers" to the radiation during the latent period, was an adaptation to phage biology of "target" theory drawn from radiation physics. They established the survival curves of the infected centers at each of several intervals of time following the infection—that is, the proportion of infected bacteria that were still able to release phage progeny after being irradiated as a function of the dose of radiation given. According to target theory, if a "single hit" is able to destroy the infective center, the survival curve should be exponential, and a plot of the logarithm of the proportion of survivors against the dose should give a straight line. If "multiple hits" are necessary to knock out the infective center, then the curve plotted in this way should at low doses be concave downward, becoming at higher doses a straight line with the same slope as the single hit curve. Applying this method to *E. coli* strain B infected by bacteriophage T2, Luria and Laterjet found that in the early stages, the curves were "single-hit," as would be expected if a single virus particle were present. In cases where they produced multiple infections by increasing the ratio of phage to bacteria, they attained some concave downward curves suggestive of "multiple hits," but in the experiments in which each bacterium was assumed to be infected by a single phage, the curves became concave upward. Luria and Laterjet were unable to interpret these results in a manner that would clarify the intracellular growth process (Luria and Laterjet 1947).

At Oak Ridge, Benzer made several improvements in what had by then become well known in the phage community as the "Luria–Laterjet experiment." To satisfy as best he could the requirement that for accurate results "growth start almost simultaneously in all cells," Benzer transferred the bacteria from their growth medium into a buffer solution in which they would exhaust their nutrient before he added the suspension of phage particles. Under these conditions the phage were adsorbed, but there was no lysis or liberation of phage. Then he started the growth at a predetermined moment by adding broth to the medium. To fulfill the second requirement for accuracy, that growth be "halted during irradiation," he pipetted samples drawn during the latent period into a buffer solution chilled in an ice bath. He could control both the start and the stop of growth to "within a few seconds," and then irradiate the samples "at leisure." In this way he not only could increase the accuracy of the results, but increase the number of different doses used and, therefore, the number of points from which to plot his survival curves. Where Luria and Laterjet had been able to obtain only three points, Benzer managed six points for each curve. With such refinements, he hoped that he would be able to obtain curves defined accurately enough to determine the number of phage particles present at different stages within the intracellular growth period. To do so one could compare the shape of the experimental curves with theoretical curves in which survival was plotted against dose according to the equation

$$y = 1 - (1-e^{-D})^n,$$

where D is the dose, and n, the number of independent targets. That the theory came "straight out of physics" was one of the factors that drew him to this problem and that eased his transition from the discipline in which he had been trained to the one into which he moved (Benzer 1952; Benzer/Holmes, interview, January 3, 2001).

Benzer began his experiments with the rapid lysis mutant of phage T2 (T2r), and *E. coli* strain B. The first results were "completely different" from what he had expected. During the first six minutes, the curves approximated those expected for single hits, but at later periods, instead of acquiring a somewhat larger "shoulder" with the same descending slope, as would be anticipated if the number of targets were increasing, the curves remained straight, with diminished slopes, indicating that the resistance to the radiation was increasing with time. The situation seemed to Benzer "pretty weird" (Benzer 1952: 66; Benzer/Holmes, interview, January 3, 2001). Despite this puzzling start, he reported to the head of his Department of Physics at Purdue in January, 1949, that "I have really fallen in love with phage and want to devote a couple of years to it, at least."[4]

After one year at Oak Ridge, Benzer transferred the remainder of his two-year fellowship to Caltech, where he worked in the laboratory of Max Delbrück. Besides experiencing directly the charismatic presence of Delbrück, who had been a legend to him ever since he had read *What is Life*, Benzer continued associations with Gunther Stent and Jim Watson begun at Cold Spring Harbor, and got to know, among others, Renato Dulbecco and Élie Wollman. He worked in the same room with Jean Weigle, with whom he carried out some phage experiments. He went along on the fabled camping trips into the desert. In March, 1950, Delbrück hosted a conference on viruses to which leading plant and animal, as well as phage virologists came. After the papers had been presented and discussed in formal sessions, some of the participants continued talking together for three more days "around camp fires in Death Valley." Benzer and the other local phage biologists put together for the occasion a "Syllabus on Procedures, Facts, and Interpretations in Phage," which summarized the methods and lore necessary to do phage research. In the summer of 1950, he attended the course on microbiology given at the Stanford Marine Biological Station by Van Neel, where he learned much about the practical problems of culturing and performing experiments with bacteria. In short, he became during his time at Caltech fully assimilated into the methods, the ethos, and the culture of the "phage group" (Delbrück 1950:[3], 100–46; Benzer 1966: 158; Benzer/Holmes, interview, January 3, 2001).

Benzer also continued at Caltech the Luria–Laterjet experiments begun at Oak Ridge. Shifting from phage T2 (and T2r, which gave similar results), to T7, he was now able to obtain a result that "resembles that predicted by target theory." As time went on through the latent period, the curves became "multiple-target in character, the average multiplicity continually increasing, while the final slope changes only slightly." With T7, therefore, he had attained the "original intent of Luria and Laterjet's experiment," but the anomalous result with T2r now seemed to him of "far greater interest." To explain the increasing resistance and retention of single-hit character shown by the curves, he suggested that a T2r phage particle "must undergo a series of successive steps $A \rightarrow B \rightarrow C \rightarrow D \rightarrow$ etc. in the course of reproduction." Each of these steps can be blocked by ultraviolet radiation, so that a single hit inactivates the phage. As the development proceeds, however, the steps that have already been passed are no longer needed, so the "cross-section" of the target decreases, accounting for the increasing resistance (Benzer 1952: 68–71).

Delbrück thought highly of Benzer's research at Caltech. Despite the fact that it was essentially an extension of the Luria–Laterjet experiment, he saw it as a "new approach to the study of intra-cellular phage growth," and judged the results Benzer had obtained at Caltech to be "very valuable." Benzer himself,

Delbrück wrote in support of an application to extend his fellowship, "is imaginative in his approach to biological problems and in the conception and design of experiments. He is painstaking in the execution of the experiments and careful and critical in their evaluation." He had also "taken every opportunity to widen his background in biology and acquaint himself with other lines of research going on at the Institute."[5] Benzer's approach to intracellular phage growth produced his first major publication, but its success proved in the long run to be more limited than it seemed at the time both to him and to Delbrück. Looking back on it several years later, Delbrück wrote that

> The intent of this work was to learn something about the replication of phages inside their host cells by studying the changes in radiation sensitivity of an infected bacterium during the course of infection. This work, though of very high quality, did not yield the desired information, due to basic aspects of replication which we still do not understand.[6]

III

In September, 1950, André Lwoff visited Caltech and took part in one of the frequent phage group camping trips, during which he invited Benzer to spend a year in his laboratory at the Institut Pasteur in Paris. As Benzer explained to his patient department head at Purdue, Lwoff had asked him to come because his experience with ultraviolet radiation was applicable to the further study of lysogeny, a phenomenon of "fundamental importance to all of biology" that Lwoff had recently discovered. This discovery, Benzer added, "indicates a close relation between the phage and the genetic material of the bacteria, suggests the possible origin of phages, and promises to clear up certain mysterious problems in modern microbiology."[7] The Committee on Growth of the National Academy of Science granted an extension of the fellowship that had supported Benzer's second year at Caltech, and Purdue generously extended his leave of absence into its fourth year (Benzer 1966: 159).[8] He and his family arrived in Paris in September, 1951. At the Institut Pasteur he shared a laboratory room with François Jacob. There he took up a research problem prompted not by Lwoff's work on lysogeny, as he had expected, but by a recently published review of "Enzymatic Adaptation in Bacteria" by Roger Stanier. While discussing the induction of the synthesis of enzymes in microorganisms by the substrates on which they act, Stanier noted that "If the rate of specific [enzyme] activity during adaptation is plotted as a function of time, the curve obtained usually has a very characteristic sigmoid form." However, he added a little further on, "the argument that a sigmoid curve would be obtained if the time required for adaptation by individual cells in the population varied according to a normal frequency distribution is sound, and almost impossible to test experimentally" (Stanier 1951: 38). The reason for his skeptical last phrase was that the kinetics of adaptation had been studied by measuring the rates of enzymatic reactions by way of the rates of formation of some immediate product, or the overall rates of gaseous exchange, which gave only the average rate for the whole population of the microrganisms examined.

When he read Stanier's statement, Benzer thought that he could challenge the impossibility of determining within a bacterial population the distributions of the time required for adaptation. Having found that in starved bacteria the intracellular development of phage is arrested, he believed he could test whether adaptation is simultaneous throughout a bacterial culture or varies from organism to organism, by making their metabolism dependent on an inducible enzyme. For someone located at the Institut Pasteur, the most obvious enzyme to test was galactosidase, inducible in *E. coli* by growing them in a medium containing lactose as their only source of carbon, a system intensively studied there for many years by Jacques Monod and Élie Wollman. At first Benzer thought he would be able to detect heterogeneity in the enzyme levels at a given time by means of the Luria–Laterjet experiment, where it would show up as a dispersal of the sensitivity of infective centers to ultraviolet radiation. Soon, however, he came upon a more direct way based on a recent discovery by Monod and Wollman that infection of their bacterial strains by phage halted the synthesis of galactosidase, and when the bacteria lysed, the enzyme was released into the medium. Consequently, if lysis were only possible in cells that continued to metabolize after they were infected by phage, and if those cells which contained more enzyme lysed sooner, then by measuring the activity of the enzymes in solution at various stages during the lysis of the culture, it might be possible to analyze quantitatively the different amounts of enzyme contained in the cells of the culture (Benzer 1953: 383–4; 1966: 159–60).

The number of cells lysed being proportional to the decrease in optical density of the culture, Benzer could determine the fraction of cells lysed at successive time intervals by removing samples "at various times," chilling them and measuring their optical density. Afterward he could assay the enzymatic activity in the supernatant fluid of each sample. The shape of a curve of enzyme activity plotted against optical density reflected the distribution of enzyme within the population. An equal distribution, for example, would give a straight line. When the bacteria were grown in a medium that contained both the inducer (lactose), and other nutrient sources of carbon, he found, "the population is essentially homogeneous in respect to enzyme content," but when the inducer was the only source, a few cells contained most of the enzyme at first, the culture gradually approaching homogeneity "as time goes on" (Benzer 1953). During the first months of his stay, Benzer and Jacob spoke little to each other. The American rose late, came to the laboratory at noon, and worked far into the night, whereas the Frenchman came at nine in the morning and left at seven in the evening. Gradually, however, they became "excellent friends." Beneath an impassive exterior, Jacob found Benzer to be a person of great charm and warmth. When asked a question, he often did not respond at once, but returned several days later with a carefully reasoned answer. Benzer was, in turn, happy with his circumstances in Paris. "The lab here," he wrote in March, 1952, "is really outstanding."

> The variety in background of the people working together—physical chemists, physicists, biochemists, immunologists, protozoologists, geneticists—gives the group an enormous strength. The lab is very well equipped, by virtue of American grants, is crowded to capacity, and buzzes with activity, arguments, and discussions. The two leading personalities are Lwoff, a fountainhead of biological insight and intuition, and Monod, a fountainhead of logic.[9]

During his last months in Paris, Benzer collaborated with Jacob on experiments extending Benzer's study of the survival curves of ultraviolet irradiated infective centers to bacteria infected by "temperate phage," such as phage λ. They found interesting changes in the resistance of the phage–bacteria complex over time, as well as shifts from single-hit to multiple hit characteristics, but these were not easy to interpret (Benzer and Jacob 1953; Jacob 1987: 290–1).

In the fall of 1952, Benzer finally returned to Purdue to begin the program of teaching and research in biophysics for which his fellowship years had been expected to prepare him. One of his first priorities was to equip his laboratory "for doing phage experiments." He felt there, at first, very alone. "I beg you to send me phage and other news," he wrote Delbrück on September 26. "After Paris the local phage isolation is almost unbearable." A little later he sent Delbrück the manuscript for a paper he planned to publish on his work in Paris on enzymatic adaptation.[10]

One month later, after traveling to the University of Indiana to talk with Salvador Luria, Benzer was feeling much better. "After visiting Luria," he wrote Delbrück,

> my lonely feeling is cured, since it is clear now that I can get a good argument at any time only 2 hours drive away—not only with Lu and his crowd, but with Sol Spiegelman and his. After considerable plumbing, dickering, and dishwashing, I have finally arrived at the stage of having plaques to count.[11]

At Purdue Benzer continued the lines of research he had begun at Oak Ridge, Caltech, and Paris. In January, 1953, he reported to Delbrück that

> I have been working on the intracellular "development" of T2 inactivated *before* infection, doing tricks with phR in the presence of cyanide—while waiting for crucial chemicals to arrive from Paris, so that I can take up the problem of the number of enzyme-synthesizing centers per cell for galactosidase (i.e. are genes the things?).[12]

Among the phage workers around the world, Benzer's reputation was steadily growing. In October, 1953 Alfred Hershey, one of the foremost phage biologists outside of the immediate Delbrück circle, wrote him, "I have just read your paper on lactose adaptation, really for the first time. What conception, what execution, what exposition (this should be in French)! Salut!"[13]

At this point in his trajectory from physics into biology, therefore, Benzer had managed in five years to advance from a beginner in the field to highly respected investigator. He had done so, not by bringing a new perspective from physics into his adopted field, for he found that approach already established in the work of those who had preceded him into phage work. His own background made him comfortable within a group that thought very much as he did. He found his research problems close at hand, in the publications of those who had recently established the field and in personal interactions with them. He engaged in short-term collaborations, such as those with Weigle and Jacob, but he preferred mainly to work alone, getting much done late at night when no one else was around. But that did not mean, as Delbrück noted,

> . . . that he does not welcome discussion of his "work-in-progress" with others, or that he closes his eyes to the research of others. On the contrary, he knows how to get the greatest good from the advice of others and his influence in any laboratory is most beneficial. It merely means that he prefers to plan and execute his own work completely and be responsible for every phase of it.[14]

We may add, that it was his careful planning, masterful execution, and lucid analysis of experiments patterned very much on those that were already known in the field that distinguished the young investigator. He had quickly integrated himself into an international network whose most active centers—Cold Spring Harbor, Caltech, the Institut Pasteur, and Luria's group at Indiana—were all familiar territory to him. There is no evidence that he brought with him an outsider's viewpoint or a vision of future developments that transcended what he experienced within these networks.

IV

Benzer has given a compelling account of the cluster of chance events that caused him to undertake the rII gene-mapping project. In preparation for a seminar on the "size of the gene," he had read a review by Guido Pontecorvo,[15] which stressed that "the various definitions of the gene were not necessarily equivalent and that high resolution genetic mapping would be required to distinguish them." At about the same time he had prepared some stocks of an r mutant of T2 phage to perform the Hershey–Chase experiment, using genetic markers to try to show that the phage genome is injected sequentially, as Jacob and Wollman had shown for bacterial conjugation. Meanwhile, to test a possibility raised by reading George Streisinger's thesis, that r mutants yield titers as high as wild types on certain strains of *E. coli*, he had plated out some T2r and T2r$^+$ on some strains he happened to have in the laboratory. For his phage class he was also preparing a lysogeny experiment for which he was growing cultures of the lysogenic bacterial strain K12(λ) and a non-lysogenic strain that had been derived from it. On the latter, both the mutant and wild-type phage made small fuzzy plaques, but on K12(λ) the plate to which the mutant had been added gave no plaques. At first he thought that he had forgot to add the phage, but when he repeated the experiment, he got the same result. "To me," Benzer wrote in 1965, "the significance of this result was now obvious at once" (Benzer 1966: 161).

> Here was a system with the features needed for high genetic resolution. Mutants could be detected by the plaque morphology using strain B. Good high-titer stocks of the r mutants could be grown using strain K12S. Strain K12(λ) could be the selective host for detecting r^+ recombinants arising in crosses between r mutants. A quick computation showed that if the phage genome were assumed to be on a long thread of DNA with uniform probability of recombination per unit length, the resolving power would be sufficient to resolve mutations even if they were located at adjacent nucleotide sites. In other words, here was a system in which one could, as Delbrück later put it, "run the genetic map into the ground." I dropped everything else and embarked on this project.
>
> (Benzer 1966: 161)

Elaborating on this story in *The Eighth Day of Creation*, Judson commented that Benzer "saw at once—there was no way to see it except instantly—that he had been presented with a flawless system for genetic mapping at very high resolution" (Judson 1979: 274). Despite Benzer's recollection, and its validation by Judson, these events and the insight that emerged from them did not happen either "at once" or "instantly." From the documents that Benzer has preserved, it is possible to confirm nearly every element in the recollected series of coincidental events that started him off on the project. Instead of taking place in the brief space of time suggested by his account, however, they occurred over a period of more than a year.

In the genetics seminar on the "Size of the gene" Benzer gave on February 20, 1953, Pontecorvo's paper was among the three references he listed. Under the heading "Nature of the gene," he described it as a "unit of *hereditary transmission, self-reproduction, mutation,* and *physiological activity*." Benzer was, in fact, discussing the question just at the time that the Hershey–Chase experiment, which he described, had persuaded his colleagues in the phage group that the genetic material was probably DNA, but two months before the appearance of the *Nature* article by James Watson and Francis Crick transformed the question of the nature of the gene. Still unaware of this impending event when he gave two lectures on the "Structure duplication" of the gene given in April, Benzer commented that "we are discussing the problem of *structure duplication* when we do not know the structure & do not know whether duplication occurs."[16]

In June, 1953, Benzer attended the Cold Spring Harbor Symposium on viruses at which James Watson presented the double helix model of the DNA molecule.

Benzer took notes on most of the lectures given during the five days. Among those from Watson's lecture was the succinct phrase that the pairing of "adenine ↔ thymine" and "guanine ↔ cytosine" "leads to suggestions as to replication." In the report on his trip that he wrote back at Purdue he included among new developments "the promising and suggestive model of Watson and Crick for the structure of DNA"; but the greatest value of the symposium for him had been the "opportunity it afforded to meet virus workers from all over the world."[17]

I have so far not located the time at which Benzer may have prepared stocks of r mutant phage to perform the Hershey–Chase experiment, or plated T2r and T2r$^+$ phage on lysogenic and non-lysogenic strains of bacteria for a classroom demonstration of lysogeny. The date at which he began systematic research on the phage mutants is, however, clearly identifiable. The first experiment in a notebook entitled "r mutants" is dated January 9, 1954—just eleven months after his seminar on the size of the gene—and the experiments continue in a nearly unbroken series until his departure for Cold Spring Harbor at the end of June. The experiments of the first ten days can be seen as preliminary ones, setting up the experimental system that included the lysogenic and non-lysogenic strains of bacteria, which enabled him to distinguish r mutants from wild-type phage. On January 20, he attempted the first crosses of two r mutants of phage T4. In a third such cross, performed on January 22 on several mutants, he classified them into those that recombined "less than 1%" of the time, "around 10%," and "20% or more."[18] He carried out two more crosses on January 23 and 25. At this point it might appear that he had, as he later remembered, "dropped everything else and embarked on the project."

For nearly three months, however, Benzer performed no more such crosses. His experiments appeared instead to lead in a different direction. He looked for other lysogenic strains, tested his mutants on various lysogenic and non-lysogenic strains, and examined the question whether lysogeny was the general characteristic of strains of bacteria on which the r mutants did not plate. He also found that some r mutants were able to plate on the same K12 (λ) strain on which the failure of other r mutants to plate had stimulated his investigation.[19] When Benzer and I reexamined these pages two weeks ago, he grew impatient with his former self for still having been, in April, occupied with what now looked to him to be mere "side issues."[20]

In those months of 1954, however, these experiments did not appear to be side issues, but the study of a different problem. On March 2, Benzer wrote to Milislav Demerec, the director of the Cold Spring Harbor Laboratory: "I am working on a problem of bacteriophage genetics (phenotypic expression of the r mutation) and would like to pass the summer at Cold Spring Harbor, especially in order to be near Hershey."[21] On the same day, he wrote to Hershey himself:

(A) Perhaps you know all this already. If not you may be interested.
The expression of the r phenotype of a (T2, T4) mutant depends upon the host. Also, some r mutations carry an associated loss of host range (in lysogenic bacteria). Thus r mutants isolated on B are usually types I and II, while r

mutants isolated on K12S or K12S(λ) (much more so than on B since types I and II do not show as r) are type III. Type II adsorbs on . . . K12 S(λ) but only a small fraction of the cells yield. The ability of the host to discriminate between types I and II goes with the presence of carried λ. All the type II mutants so far tested are genetically distinct and I think this system may lead to a sort of "physiological" genetics of phage.

(B) I would like to know whether these classes are associated with different locations on the chromosome map. Rather than map all these mutants, I would be most grateful for samples of your already-mapped T2 mutants (as many as you are willing to give) and double mutants.[22]

Finally, Benzer asked if he might work in Hershey's lab during the summer. In a third letter written the same day, Benzer described his project in a somewhat different way in a request to O.M. Ray of the National Research Council for permission to use funds from his grant to attend the International Photobiology Congress in Amsterdam in August:

I am working on a problem of interference between carried and infective bacterial viruses. While wild type (T2 and T4) can multiply in a lysogenic or on a non-lysogenic host, certain mutants can multiply only in the non-lysogenic form. These mutants will adsorb to and kill the lysogenic cell, but, due to the presence of the carried virus, the process of development is blocked at some unknown stage. Dr. F. Jacob (Paris), one of the foremost workers on lysogenic bacteria, has been working on closely related problems and a consultation with him would be invaluable.[23]

On March 31, Frank Lannie wrote Benzer that

When we visited Urbana a few weeks ago, we heard in a rather vague way about some interesting experiments you have been doing with T2r$^+$ and T2r in relation to host range. Nobody seemed to know the details in reliable fashion. If it is not too much trouble, would you care to send us a brief account.[24]

The events followed so far suggest that, even if we accept Benzer's recollection twelve years later as an accurate description of the insight that launched him on his investigation of the r mutants, it does not necessarily follow that this remarkable, fortuitous sequence of encounters "presented" Benzer with a clear vision of the directions he would take during the next several years. Hershey had begun mapping phage T2 in 1948, and Gus Doerman had mapped eight mutants of T4, including several r mutants, in 1952 (See Holmes 2000: 120–2). The idea of extending these maps further was obvious enough so that others were already interested in doing that. What was "obvious at once" to Benzer may have been only that he had the opportunity to carry these maps down to finer levels of resolution. Even so, that prospect was evidently not at first powerful enough to dominate the early stages of his project, or to prevent him from redefining it for a time around the host-range problem that was also implicit in the initial chance observations.

Rather than to use the failure of r mutants to plate on a lysogenic bacterial strain "at once" as a tool for mapping, he turned to investigate the phenomenon itself. That he did so is easy to explain. Not only was Jacob interested in lysogeny, but Benzer had himself collaborated with Jacob in experiments designed to elucidate the intracellular growth period of temperate phage. That something in lysogenic bacteria seemed to "block" the growth of the T2r mutant thus provided the possibility to learn more about a problem in which he had previously been involved.

Why, then, did this interlude disappear from his remembered account? It is possible that he simply omitted it to keep his story simple, but equally likely that as his gene-mapping project built momentum, this episode faded from his memory, because it had no necessary connection to what came after.

That Benzer did begin to map his r mutants early in May may owe in part to another fortuitous event that did not remain in his remembered "Eureka" experience. During March he had requested from the two phage biologists who had previously been mapping phage mutants, some of their mutant stocks. On April 3 one of them, Gus Doermann, wrote him:

> The markers that I have mapped are all given on the map in the CSH symp[osium] paper. Sending you my stocks, however, has one condition. This arises from the fact that everyone wants to use genetically known material, but no one is willing to do the more or less thankless and dull job of mapping the markers. Therefore the condition is that you must promise to locate on the T4 map at least two of your independently arising mutants.[25]

Whether motivated by Doermann's goad, or returning to a plan he had already entertained, Benzer took up the mapping project in earnest on May 4, and pursued it intently for the rest of May and June. His progress was rapid. By May 21, he had already produced a map summarizing the results of 22 crosses, which was essentially the same as the one he published in his first paper on the topic eleven months later.[26]

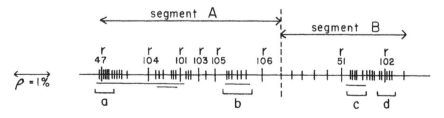

Figure 2.1 Preliminary locations of various rII mutants, based upon spot tests (Benzer 1955). Reproduced with kind permission by the author.

On May 28 Benzer wrote down in his research notebook a page of "thoughts on the gene." He began,

> It occurs to me that the "r" region under study can be interpreted as a *single gene*. The length of the region corresponds to the order of 10,000 nucleotides per strand. If one considers that each nucleotide pair determines a particular amino acid, then the region could correspond to the synthesis of a polypeptide chain 10,000 amino acids long.[27]

Obviously Benzer had at this point not thought through what soon became known as the "coding problem," or it would have occurred to him that four nucleotides taken singly could not determine twenty different amino acids. That he had not done so, however, testifies to the incipient nature of the rest of his insight. After estimating the molecular weight of the DNA chain and considering the reversion rates of mutations at particular nucleotides, Benzer continued,

> This case gives a clear distinction between the gene as a unit of recombination or mutation and as a unit of physiological action. The nucleotide (or a very small number of them) is probably the unit of recombination or mutation, while the whole string of nucleotides (10,000 of them) is the unit of physiological action. An upper limit on the no. of nucleotides per unit of recombination can be obtained by isolating many more r mutants and seeing at what point "allelic" ones start to occur ... It would be nice to be able to isolate the specific protein involved & to see if its properties are different for the r mutants and the wild type.[28]

This document has the feeling that Benzer was putting down insights just as they came to mind. Though we cannot be sure that some elements of what he wrote here had not already occurred to him before, so that these passages may represent an integration of thoughts that had previously appeared to him in more fragmentary or weaker forms, it seems clear that he felt, on May 28, 1954, that these were new thoughts. They contain much of what had been, according to his later recollection, an insight that he had acquired in a rush of coincident events beginning with the reading of Pontecorvo and ending with the realization that he had in hand a system with which he could "run the genetic map into the ground." How can we explain that he compressed into such a cluster of events and developments a period of time that was actually spread out over fifteen and a half months?

As so often happens, Benzer's later memory seems to have been in part reshaped by the outcome of his "project." Howard Gruber has argued, and illustrated by following the early thought of Charles Darwin on the species question, that creative scientists often in retrospect unconsciously collapse into a single "Eureka" moment the origin of a decisive new insight that can be shown, if the contemporary record of their thoughts is dense enough, to have been acquired more gradually, by cumulative smaller flashes of insight (Gruber 1981). Benzer's account of the origins of his project is a remarkable and full example of this process.

That the mapping project may have arisen in May as a diversion from the host-range project that he had been pursuing in February and March, rather than the host-range project as a temporary diversion from a mapping project conceived in January, is suggested by a letter that Benzer wrote to Gunther Stent from Cold Spring Harbor on July 17:

> I have gotten into an offshoot of the r problem. The applications of the effect are more interesting than the phenomenon itself. It is technically possible to study, by genetic means, the structure of the r "gene" down to the individual nucleotides, since I can detect recombination frequencies that small. So I am going to be isolating and mapping hundreds of r mutants next year.[29]

At the time he wrote Stent, Benzer had mapped eight T4 rII mutants selected from a larger number of crosses. All had the "same phenotypic effect of interfering with the ability of the phage to plate on K12 (λ)," although in some of the mutants, which he called "leakers," the effect was not total. That he could "detect recombination frequencies down to the individual nucleotides" was a prediction, rather than an achievement already reached. He had calculated that from the length of the phage DNA and the total number of nucleotide pairs contained in it according to Hershey, the frequency of recombination between "markers separated by a single nucleotide pair . . . is not less than 5×10^{-6}." The method he had in hand, he estimated, "allows the detection of wild-type recombinants in proportions as low as 10^{-8}." Consequently, it seemed "feasible to establish this fundamental frequency experimentally" (Benzer [1954]).

At Cold Spring Harbor, Benzer gave a seminar on his new project, discussed it intensively with others there, and wrote a paper describing his results and their implications. One copy he sent to Delbrück, who was, at the time, in Göttingen. On July 27, Delbrück wrote back,

> Dear Seymour. I think your paper is tremendously courageous, in fact, I think you must have had a triple highball before writing it. I object to publishing it in the present form because two thirds of it may be proved wrong by your own experiments (or may be vindicated) even before the paper is out. In this situation, if you don't want to be patient enough to do the experiments which will vindicate or disprove the two basic ideas, viz.
> (a)   that c[ross] o[ver] can occur between any pair of nucleotides
> (b)   that mutation is a substitution of a single false base . . .[30]

Delbrück suggested that Benzer write a paper of half the length saying that he had a dandy system to study closely linked loci, clustered in what appear to be pseudo-alleles divisible by mutation, and that a rough estimate gives 16,000 nucleotides for this "gene."[31]

Benzer's reply is not recorded, and it probably took place in the form of a conversation with Delbrück in Amsterdam. Benzer also sent a copy to Edwin Lennox, who read the paper at the Phage Meeting at Cold Spring Harbor, on August 25, while Benzer was in Europe, and to Sidney Brenner, who made an abstract of it that appeared in the *Phage Information Service* bulletin, an informal mimeographed summary of papers presented at recent meetings, and other news, that Delbrück circulated among the members of the phage group. The contents of the abstract confirm that Benzer had not yet achieved cross-over frequencies so low as to approach the dimensions of single nucleotides, but only inferred from his initial results and calculations that "extremely fine mapping of a single gene is possible" (Benzer [1954]: 4).

"From the experimental data," Benzer wrote, "it is possible to estimate the length of DNA in the rII gene." Making various assumptions, he concluded that the upper limit for the rII gene was 1.6×10^4 nucleotide pairs," while the lower limit was 32 pairs. "This rather broad range for the size of the rII gene," he predicted, "will narrow rapidly as more mutations are found." It is notable that Benzer tacitly assumed that the region of the phage DNA in which the rII mutants mapped constituted one gene, apparently because all of the mutants were characterized by the same functional deficit, "a failure of the phages to form plaques" on strains of *E. coli* that carry the prophage λ (Benzer [1954]: 4).

The fact that he found the frequency of reversion of rII mutants to wild type to differ from the frequency of forward mutations led Benzer to a critique of the "Watson–Crick model of mutation due to tautomerization of the bases." The prediction of the model that "the type of changes in forward and reverse mutations should be identical . . . was not borne out by the data." To account for this discrepancy, Benzer suggested that the tautomerization of a base "can be influenced by neighboring bases and hence a wide range of mutation frequencies would be expected" (Benzer [1954]: 5).

On October 30, Benzer attended an informal research conference at the University of Indiana "held for the purpose of free interchange of latest results among midwesterners engaged in research on . . . closely related subjects." In his report of the trip, he wrote that "I described my work on the fine structure of a bacteriophage gene, for which it has been possible to demonstrate that while crossing over within the gene is readily possible, there is a limit at which one runs into atomicity, i.e. there are sub-units within which crossing over cannot occur. The size of the subunit can be estimated in molecular terms and amounts to no more than a few hundred nucleotide units (the molecules of which genes are made)."[32]

His report suggests that Benzer had heeded Delbrück's admonition to be more cautious, and was limiting his claims to what his current results demonstrated, rather than to what his calculations suggested that further results would ultimately show. Meanwhile, the abstract of his paper in the *Phage Information Service* was already attracting attention among the inner circle of the phage group who received copies. On December 22, Hershey wrote to Benzer,

> If I can judge from Brenner's abstract, I have one minor suggestion. The word gene is full of confused notions among geneticists. ... You are introducing a new (to phage) and satisfactory definition: that region of the phage in which mutations produce a given class of effects. This, however, disagrees in this instance with all previous definitions (see concluding remarks in my paper in Advances in Genetics). In that paper I proposed that when such disagreement arose, the word gene should be abandoned for its descriptive equivalents—in your case some sort of "action unit." I still think this is a good idea. At the same time we should stop using the word gene in any of its other senses.[33]

In reply, Benzer wrote on December 30,

> I agree with you about "genes." In fact I plan to use that dirty word only once—in order to state that I am not going to use it. At the time of the first draft, I did not realize how bad the situation was, but since then I have talked to geneticists and no two seem to agree. How do you like "functionally related region of a linkage group?" This is a jaw-breaker, but almost always can be abbreviated by "region." I like region better than locus, because the geneticists are just as confused about locus as about gene. Pontecorvo, with whom I have discussed this question, likes region also. However, there remains the difficulty of defining "functionally related" without specifying the function. The specification of a particular amino acid is a function, as is the specification of an entire enzyme.[34]

After examining these letters with me in January, Benzer agreed that it was the exchanges included or mentioned in them that aroused his awareness of the need to substitute new terms for each of the properties previously associated with the ill-defined gene. He did not, however, immediately give up that much abused word, but continued to discuss the various units of the gene in his lectures up until April 1956. Then, in the notes for his next presentation of the subject, at the Mcollum Pratt Symposium on the Chemical Basis of Heredity in Baltimore in June, he wrote: "significant—word gene missing from title." After discussing the "difficulty in terminology" which he ascribed to "a hangover from the dismal gene," he asserted that "we should say what we mean," and introduced the three terms "recon," "muton" and "cistron," with which he hoped to replace the venerable term in which these three distinct genetic units had previously been merged.[35] Thus, as in the origin of his mapping project, Benzer's decision to avoid the gene was not a sudden one, but the outcome of a gradual evolution, extending in this case over nearly two years.

During the same month in 1954 that Hershey alerted him to the troubles with the gene, George Streisinger wrote Benzer from the mecca of phage biology at Caltech:

> We have been overwhelmed with your experiments—the idea is really beautiful.
> What is the latest poop? How close can you get two markers? Inspired by you
> I have been thinking of starting some experiments along the same lines with
> the h locus.[36]

That the experiments were not only beautiful, but exciting and potentially
profound, was a viewpoint that was by now spreading through the phage group
on both sides of the Atlantic. Benzer was emerging from the ranks of promising
young phage workers to the status of one of its rising stars.

V

I have presented Seymour Benzer entering molecular biology, both before and
after he began the "adventure in the rII region" that cast him, several years later
in the role of master architect of the bridge between molecular biology and classical genetics, as someone proceeding one step at a time along an investigative
pathway that took its direction from the local circumstances in which he found
himself at each step, that changed directions as new opportunities came into view,
and as new directions sometimes crowded out still promising prior lines of investigation not yet completed. Benzer learned from those around him, shared their
methods and their outlooks. He did not seek a loner's visionary path. The experimental system that fell into his hands through a fortunate cluster of chance circumstances that would lead him to a position in which the former physicist who
had never taken a course in classical genetics would be seen as the one who
brought that field into the molecular age was not something that he could foresee
from the beginning. Like all scientists who chart a bold course into the unknown,
he could not tell, when he set out on it, "how it is going to turn out."

Notes

1 S. Benzer: Notes on Phage Course: June 28–July 17, 1948 at Cold Spring Harbor, L.I.,
   unpaginated, Benzer personal files.
2 Ibid.
3 Abbreviated as: Benzer/Holmes, interview, January 3, 2001.
4 S. Benzer to K. Lark-Horovitz, January 24, 1949, Benzer personal files.
5 M. Delbrück to Ch.E. Richards, November 3, 1950, Delbrück Papers, Caltech
   Archives, 2.24.
6 M. Delbrück to B.D. Davis, August 12, 1957, Delbrück Papers, Caltech Archives, 2.24.
7 S. Benzer to K. Lark-Horovitz, October 2, 1950, Benzer personal files.
8 M. Delbrück to Ch.E. Richards, November 3, 1950.
9 S. Benzer to K. Lark-Horovitz, March 7, 1952, p. 2, Benzer personal files.
10 S. Benzer to M. Delbrück, September 26, 1952, Delbrück Papers, 2.24.
11 S. Benzer to M. Delbrück, October 27, 1952.
12 S. Benzer to M. Delbrück, January 23, 1953.
13 A. Hershey to S. Benzer, undated [October 15, 1953], Benzer files.

14 M. Delbrück to B.D. Davis, August 12, 1957; Benzer/Holmes, interview, January 3, 2001.
15 For a discussion of Pontecorvo's views and the extent to which they forshadowed Benzer's mapping project, see Holmes (2000: 118–20, 123–4).
16 "Genetics Seminar, Purdue, February 20, 1953," "Bio 156 lecture—April 10, 1953," p. 2, Benzer personal files.
17 Benzer, Notes on Cold Spring Harbor Symposium, June 5–11, 1953; "Report of Trip to Virus Symposium," Benzer files.
18 Research notebook "r mutants", n.1, January–March 1954, January 9 to January 23, Benzer personal files.
19 Research notebook "r mutants", n. 1, January 27 to April 14, 1954, Benzer personal files.
20 Benzer/Holmes, May 2, 2001.
21 S. Benzer to M. Demerec, March 2, 1954, Benzer files.
22 S. Benzer to A. Hershey, March 2, 1954, (rough draft), Benzer files.
23 S. Benzer to O.M. Ray, March 2, 1954, Benzer files.
24 F. Lannie to S. Benzer, March 31, 1954, Benzer files.
25 A.G. Doermann to S. Benzer, April 3, 1954, inserted in research notebook no.1.
26 Research notebook "r mutants," no.1, May 4 to May 21, 1954.
27 Research notebook "r mutants," no.1, May 28, 1954.
28 Ibid.
29 S. Benzer to G. Stent, July 17, 1954, Stent Collection, University of California Archive, Berkeley.
30 M. Delbrück to S. Benzer, July 27, 1954, Benzer files.
31 Ibid.
32 "Research conference at University of Illinois—S. Benzer," Benzer files.
33 A. Hershey to S. Benzer, December 22, 1954.
34 S. Benzer to A. Hershey, December 30, 1954, Benzer files.
35 "Baltimore lecture, June 19, 1956," Benzer personal files.
36 G. Streisinger to S. Benzer, December 5, 1954, Benzer files.

Bibliography

Benzer, S. (1952) "Resistance to Ultraviolet Light as an Index to the Reproduction of Bacteriophage," *Journal of Bacteriology*, 63: 59–72.
—— (1953) "Induced Synthesis of Enzymes in Bacteria Analyzed at the Cellular Level," *Biochimica et Biophysica Acta*, 11: 383–95.
—— (1954) "The Fine Structure of a Gene in Bacteriophage," *Phage Information Service*, Bulletin No. 8: 4–5.
—— (1955) "Fine Structure of a Genetic Region in Bacteriophage," *Proceedings of the National Academy of Sciences of the United States of America*, 41: 344–54.
—— (1966) "Adventures in the rII Region," in J. Cairns, G.S. Stent, and J.D. Watson (eds) *Phage and the Origins of Molecular Biology*, Cold Spring Harbor: Cold Spring Harbor Laboratory.
Benzer, S. and Jacob, F. (1953) "Etude du développement du bactériophage au moyen d'irradiations par la lumière ultra-violette," *Annales de l'Institut Pasteur*, 84: 186–204.
Delbrück, M. (ed.) (1950) *Viruses 1950*, Pasadena: California Institute of Technology.
Gruber, H.E. (1981) *Darwin on Man: A Psychological Study of Scientific Creativity*, 2nd edn, Chicago: University of Chicago Press.
Holmes, F.L. (2000) "Seymour Benzer and the Definition of the Gene," in R. Falk, P. Beurton, and H.-J. Rheinberger (eds) *The Concept of the Gene in Development and Evolution: Historical and Epistemological Perspectives*, Cambridge: Cambridge University Press.
Jacob, F. (1987) *La statue intérieure*, Paris: Seuil.

Judson, H.F. (1979) *The Eighth Day of Creation: The Makers of the Revolution in Biology*, New York: Simon and Schuster.

Luria, S.E. and Laterjet, R. (1947) "Ultraviolet Irradiation of Bacteriophage during Intracellular Growth," *Journal of Bacteriology*, 53: 149–63.

Rheinberger, H.-J. (1997) *Toward a History of Epistemic Things: Synthesizing Proteins in a Test Tube*, Stanford: Stanford University Press.

Schrödinger, E. (1946) *What is Life: The Physical Aspect of the Living Cell*, Cambridge: Cambridge University Press.

Stanier, R. (1951) "Enzymatic Adaptation in Bacteria," *Annual Review of Microbiology*, 5: 35–56.

Weiner, J. (1999) *Time, Love, Memory: A Great Biologist and His Quest for the Origins of Behavior*, New York: Alfred Knopf.

3   Walking on the chromosome

Drosophila and the molecularization of development

Marcel Weber

While the fruit fly *Drosophila melanogaster* played a crucial role in the establishment of genetics as a science, the molecularization of genetics in the 1950s and 1960s was largely a result of work involving microorganisms, especially the colon bacillus *Escherichia coli* and its bacteriophages (see Chapter 2 by Holmes, this volume). However, beginning in the 1970s, *Drosophila* came back on the scene as one of the major laboratory organisms of molecular biology, especially for research on the molecular basis of development. For example, in 1984, molecular work on *Drosophila* led to the discovery of the homeobox, an evolutionarily highly conserved DNA sequence element with a length of 180 base pairs that is found in eukaryotic organisms from yeast to humans. The homeobox codes for a DNA-binding domain, that is, a part of a protein molecule that makes this molecule bind to DNA in a specific manner. Genes containing a homeobox often have a gene regulatory function. In insects, such genes seem to be involved in important developmental decisions, for example, the decision as to which body segments different parts of the embryo should develop into. Some homeobox genes perform a similar function in vertebrates. Even though the various roles of different homeobox genes are only beginning to be understood, there can be no doubt that the discovery of the homeobox was a crucial event in the molecularization of developmental biology (Gehring 1985, 1998).

The homeobox and a series of other discoveries are the reason why, today, *Drosophila* is considered to be one of the most important model organisms for trying to make sense of the human genome sequence (Adams *et al.* 2000; Kornberg and Krasnow 2000; Rubin and Lewis 2000). The fly was only the second multicellular organism for which the full genomic DNA sequence was determined. Indeed, this tiny insect has come a very long way since Thomas Hunt Morgan spotted the first *Drosophila* mutants in the famous "fly room" at Columbia University in 1910. For this reason, *Drosophila*'s history as a laboratory organism provides a unique opportunity to examine the changing roles of mapping in twentieth-century genetics.

In recent years, an increasing number of historians and philosophers of biology have directed their attention to the role of experimental systems and model organisms in biological research.[1] It is widely agreed that the impact of such systems on the development of scientific disciplines is profound. The reason, according to most historians, is not just that such systems help to *solve* research problems. Rather, such systems often also play an important role in *defining* research problems in the first place. Some have suggested that model systems and the scientific problems they help to solve are co-constructed (Clarke and Fujimura 1992; Lederman and Burian 1993).

Robert Kohler has shown that *Drosophila* entered the laboratory for apparently extraneous reasons, but turned out to be extremely well adapted and adaptable to life in the laboratory (Kohler 1994). *Drosophila*'s capacity to produce an apparently endless variety of mutants with visible phenotypic effects placed stringent imperatives on the organization of laboratory practice. In Kohler's view, the invention of genetic mapping by Morgan and his students is best viewed as a re-organization of laboratory practice that enabled them to cope with *Drosophila*'s enormous fecundity as a breeder-reactor for new mutants. I have argued elsewhere that classical mapping techniques also played an *epistemic* role, namely, they served to represent genetic structures and fine structures (Weber 1998). Furthermore, these techniques played an important role in the molecularization of microbial genetics (Weber 1998). In this chapter, I want to examine the role that the mapping techniques derived from classical *Drosophila* genetics played in the molecularization of developmental biology in the 1970s and 1980s.

In the following section, I tell the story of the cloning of the first fly genes, which was accomplished soon after recombinant DNA technology became available in the 1970s. I place particular emphasis on the role of classical mapping techniques, especially cytological mapping of giant chromosomes, a technique that had been invented in the 1930s. Then I show that these techniques, which once had a mainly representational function, assumed a preparative role. Finally, I examine the respective roles of the classical mapping tradition and earlier work in developmental genetics in the discovery of the homeobox.

Cloning fly genes

Our story begins in the midst of the cloning revolution, which originated at the Stanford University biochemistry department, where the first *in vitro* recombined DNA molecules were produced in 1972 (Morange 1998: ch. 16). The scientific and technological promises as well as ethical concerns that the first genetically engineered organisms generated need no repeating here. In the present context, cloning[2] designates a set of methods by which DNA fragments from any organism are inserted *in vitro* into a vector—a small replicating unit (typically a bacterial plasmid or a bacteriophage)—for amplification and subsequent molecular analysis, experimental intervention, and gene transfer. It did not take long until the new recombinant DNA technology, which was originally developed on bacteria, bacteriophage, and animal viruses was applied to *Drosophila*, the genetically best-understood multicellular organism.

In fact, some of the standard methods of gene cloning were developed using DNA isolated from *Drosophila*. For example, David Hogness's laboratory at Stanford developed a method called colony hybridization, which enables the isolation of any DNA fragments that are complementary to a given RNA molecule (Grunstein and Hogness 1975). In this technique, the molecular biologists start out with a so-called genomic library. Such a library was produced in Hogness's laboratory by cutting the total DNA of *Drosophila* into thousands of small, random fragments, which were then inserted into cloning vectors. The vectors carrying *Drosophila* DNA were then transferred into bacteria and the bacteria were allowed to grow into visible colonies on petri dishes. Thus, each bacterial colony on these dishes carried a different piece of *Drosophila* DNA. By dissolving the bacterial cells, this DNA was attached to special nitrocellulose filters. Then, Hogness and his co-workers added to these filters several species of RNA molecules that they had isolated biochemically from *Drosophila* ribosomes, the so-called 18S and 28S ribosomal RNA (rRNA).[3] These rRNAs were radioactively labeled. The radioactive molecules were allowed to bind to those DNA fragments in the genomic library whose sequence was complementary to that of the rRNAs, a process that is called hybridization.[4] Thus, the genes coding for the rRNAs became detectable as radioactive spots on the filter papers. The researchers could then simply go back to the bacterial colonies that produced these signals. These colonies contained the genes that they wanted. In this way, the DNA fragments containing the rRNA genes could be picked out from the thousands of fragments present in the library.

Hogness was originally not a *Drosophila* geneticist (although he quickly became one); he was a classical molecular geneticist who had previously worked with *E. coli* and bacteriophage. However, several established *Drosophila* laboratories became interested in doing molecular work on the fly. One such laboratory was Walter Gehring's at the University of Basel, which was working on embryonic development in *Drosophila*.[5] They established a gene bank (obviously the Swiss version of a genomic library) by inserting DNA fragments produced by the mechanical shearing of total nuclear DNA into a bacterial plasmid (Gehring 1978). Then, they used the colony hybridization technique on this gene bank to isolate the *Drosophila* 5S rRNA genes (Artavanis-Tsakonas *et al.* 1977). Next, Gehring's laboratory turned to heat shock genes. Like most cells, *Drosophila* expresses a number of specific proteins after a heat shock and shuts down the expression of most other genes. Thus, after a heat shock, cells contain mostly mRNA[6] copies from heat shock genes. By isolating the mRNA from heat-shocked cells, Gehring and his co-workers obtained a specific RNA probe that they could use for a colony hybridization screen to isolate the heat shock genes (Schedl *et al.* 1978; Artavanis-Tsakonas *et al.* 1979; Moran *et al.* 1979).

The colony hybridization method requires a purified RNA molecule that is complementary to the gene being cloned. In the case of the rRNA and heat shock genes, abundant and biochemically pure RNA species were available for this task. The common feature of these first cloning attempts was that the gene products of the genes being cloned were known (rRNA and heat shock proteins, respectively). However, not all genes that interested the geneticists offered such an easy starting point. For most of *Drosophila*'s genes, the gene products were unknown at that time. For this reason, Hogness's laboratory developed an ingenious method for cloning those *Drosophila* DNA sequences about which nothing was known, except for their chromosomal location. This method came to be known as chromosomal walking.

Chromosomal walking makes full use of the powerful resources of classical *Drosophila* genetics. One of these resources is a special cytological mapping technique that had been used by *Drosophila* geneticists since the 1930s. The salivary glands of *Drosophila* larvae contain giant chromosomes that can be stained and made visible in an ordinary light microscope. Staining produces a stable pattern of chromosomal bands. Geneticists were able to assign most of the known genes of *Drosophila* to some of these bands. (This was done by simply correlating the phenotypic effects of mutant genes with the visible rearrangements of the chromosome bands in the mutants). These maps corresponded well with the maps obtained by the classical method of linkage mapping (Falk 2004), and were especially useful for mapping large chromosomal rearrangements such as inversions (Gannett and Griesemer 2004). For molecular cloning, the giant chromosome maps also offered an additional advantage, namely, they could be used for a technique called *in situ* hybridization. In this technique, a radioactively labeled DNA probe is allowed to bind to the giant chromosomes. The DNA will bind to a location that matches the sequence of the DNA probe. In this way, the location of the DNA sequences can be visualized directly on the giant chromosome under an ordinary light microscope (see Figure 3.1). This technique was especially useful for the cloning work because it allowed geneticists to know where any piece of cloned DNA belonged on the chromosome.

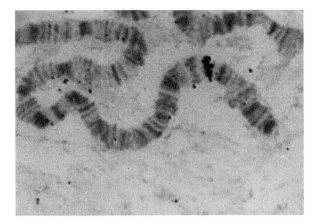

Figure 3.1 In situ hybridization of a cloned DNA fragment. Here, a DNA probe containing the *Drosophila white* gene was hybridized to a giant chromosome preparation. The location where the radioactively labeled DNA fragment bound shows up as a black band on an x-ray film. Photograph supplied by Walter J. Gehring (reprinted with permission).

Now, I turn to the technique of chromosomal walking. A chromosomal walk can start with any fragment of previously cloned *Drosophila* DNA that is located not too far from the region to be cloned. In the first step, the cytological location of the cloned DNA sequence is determined by *in situ* hybridization to giant chromosomes. In the second step, the starting DNA fragment is used to pick from a genomic library random fragments which overlap with it. The new fragments are mapped by *in situ* hybridization. Then they can be aligned to the starting fragment by restriction endonuclease mapping[7] in order to determine which fragment is farthest from the starting point in the direction of the walk. This second step is repeated many times and generates overlapping cloned DNA fragments that lie farther and farther from the starting point. The end point of a chromosomal walk is the region of interest, that is, the known or assumed location of a gene or gene complex of interest (as determined by classical cytogenetic mapping). In order to save time, it is possible to jump large distances on the chromosome by using chromosomal rearrangements such as inversions. Rearrangements such as small deletions can also be used to narrow down the position of a cloned DNA fragment. For instance, if such a fragment fails to hybridize to a giant chromosome carrying a deletion, this indicates that the cloned sequence lies within the breakpoints of the deletion.

To my knowledge, this technique of using *in situ* hybridization to clone DNA sequences for which only their cytological location is known is unique to *Drosophila*. Hogness's laboratory first used it to clone the *rosy* and *Ace* genes.[8] While this walk was in progress, the researchers learned from E.B. Lewis of inversions with endpoints in the *Bithorax* complex and in the region in which they were walking, and they used these to jump into the *Bithorax* region (Bender *et al*. 1983). Thus, rather accidentally, the first homeotic gene complex was cloned.

Homeotic mutations were first described by William Bateson as early as 1894. They are characterized by transformations of body parts into more anterior or more posterior structures. The *Drosophila* Bithorax locus harbors several such genes, mutations of which can cause bizarre phenotypes like flies with a second pair of wings. Another homeotic locus called *Antennapedia* is the site of mutations with even more bizarre phenotypes such as a transformation of the antenna on the fly's head into legs. Interest in homeotic loci stemmed from their presumed role in controlling embryonic development (see later).

Around 1980, the laboratories of Walter Gehring and Thomas Kaufman set out to clone the *Antennapedia* gene.[9] This was a strenuous effort, which, in the case of the Gehring group, took more than three years to complete (Gehring 1998: 46). I shall briefly discuss the strategy used by this laboratory.[10]

All they knew when they started their walk was that *Antennapedia* (*Antp*) mapped to the chromosome band 84B in cytological maps. Gehring and co-workers started their walk in the band 84F, from which clones containing tRNA genes were available. They then walked towards the *Antp* locus until they encountered the breakpoint of a known chromosomal inversion (*Humeral*) and used this to jump closer to the *Antp* cytogenetic map position (see Figure 3.2).

They then had to walk another 230,000 base pairs until they reached the *Antp* region. This by no means completed the task, since the cloned fragments were

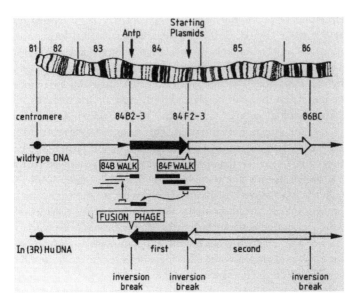

Figure 3.2 Strategy for cloning the homeotic gene *Antennapedia*. The top of the figure shows the location of the *Antennapedia*-locus and the starting sequences (which had been cloned previously) on giant chromosome preparations. The thick black arrow shows the chromosomal inversion *In(3R) Humeral*, which was used to jump into the *Antennapedia* region. Fusion phage refers to a recombinant bacteriophage carrying DNA from the inversion break point, which was used for jumping across the inverted region. Drawing supplied by Walter J. Gehring (reprinted with permission).

rather large and the *Antp* locus first had to be localized on the final clones from the walk. For this purpose, Gehring and co-workers mapped various inversions and deletions in the *Antp* region (i.e. different *Antp* alleles) by *in situ* hybridization. Furthermore, they screened a cDNA[11] library prepared from embryonic mRNA for the ability to hybridize to the cloned chromosomal region where *Antp* was mapped with the help of the inversions and deletions. In other words, a number of cDNA molecules were screened for complementarity to clones, that hybridized to the known cytogenetic location of *Antp*. They finally obtained two clones, a 2,200-base-pairs-long DNA fragment and another 3,500-base-pairs-long fragment.

The problem in gene cloning is always to know what one has cloned. This problem is especially acute if, as in the case of *Antennapedia*, absolutely nothing is known about the DNA sequences to be cloned. Garber *et al.* (1983) produced the following evidence that they had actually cloned DNA carrying sequences from the *Antennapedia* locus. First, they showed that the cDNA clones they had isolated represent RNA derived from a DNA region spanning about 100,000 base pairs, where the *Antennapedia* gene mapped. This is a very large region for a single gene (most genes are much shorter than this). But Gehring and his co-workers were able to show that all the known DNA rearrangements that affect the function of *Antennapedia* lie within this 100,000 base pairs region. This was at least good evidence that the cloned DNA fragments contained sequences necessary for *Antennapedia* function. Later, it turned out that the gene contains huge intervening sequences (introns), which explains its unusually large size.

This example shows clearly that the classical methods and concepts of *Drosophila* genetics played a crucial role in identifying molecular genes such as *Antennapedia*. Even though it was recombinant DNA technology that allowed the molecular biologists to isolate and amplify specific fragments of *Drosophila* DNA, the classical methods as well as the vast collections of different mutants at the *Bithorax* and *Antennapedia* loci were necessary in order to locate the genes sought on the cloned DNA. It was thus the combination of the enormous experimental resources from classical *Drosophila* genetics with recombinant DNA technology that made *Drosophila* such a powerful system for gene cloning.

In the following section, I will examine the function of mapping in the molecularization of *Drosophila* more closely.

From representation to preparation

The original purpose of classical mapping was the representation of genetic structures and fine structures (Weber 1998). In addition to this epistemic role, genetic maps provided a convenient way of classifying and cataloguing the thousands of *Drosophila* mutants that kept turning up in the laboratory (Kohler 1994: 54–8). The work discussed in the previous section now points us to a novel use of genetic maps. As we have seen, molecular *Drosophila* geneticists employed a combination of classical mapping techniques and recombinant DNA technology in order to isolate or clone the first fly genes. In contrast to classical genetic or cytological mapping, cloning is a *preparative* technique. The immediate goal of this experimental activity is not to produce a spatial representation, but the production of *research materials* for further study. The raw product of cloning is typically an amplified recombinant plasmid or bacteriophage that contains a fragment of specific DNA from a source organism—in our case *Drosophila* DNA. This recombinant DNA can then be used for further experimental work, for example, sequencing, *in vitro* recombination, site-directed mutagenesis, or gene transfer. A substantial part of the experimental work in molecular biology has this preparative character. After the successful completion of such preparative work, conducting an actual experiment could take just a few hours, while cloning the necessary gene took years in the early 1980s. The problem in gene cloning, of course, is to clone the right DNA sequences, that is, sequences that contain genes or genetic regions of interest. Genomic DNA libraries contain thousands or millions of different fragments. How could molecular biologists select fragments that contained a specific gene or genetic region?

I have shown in the last section that there were several techniques available for selecting DNA fragments from genomic libraries. One technique made use of RNA molecules (rRNA or mRNA) that are complementary to the sequences being cloned. But this technique was not widely applicable, since specific RNA probes for colony hybridization were only available for a few genes.[12] By contrast, the technique of chromosomal walking—which is basically also a method for selecting DNA fragments from a genomic library—can be applied to any gene for which there are phenotypically detectable mutations. This opened the possibility of cloning genes about which nothing was known except their chromosomal location (also known as positional cloning). Specifically, nothing was known about the gene *products*. For example, this was the case for the homeotic loci of the *Bithorax* and *Antennapedia* complexes. Although the specific phenotypic effects of mutations at these loci gave some clues as to their role in development (see later), absolutely nothing was known about the proteins encoded by these genes. Furthermore, these gene products and their mRNAs are present at extremely low concentrations, which made their biochemical isolation difficult, if not impossible. What made the cloning of these genes possible is the fact that their chromosomal positions had been determined to a great degree of precision by cytogenetic mapping.

The method of *in situ* hybridization in giant chromosomes allowed geneticists to navigate on the chromosome by visually localizing newly cloned DNA fragments under an ordinary microscope. The large collection of available and cytologically mapped inversion, deletion, and insertion mutants of *Drosophila* could be used to speed up the process of chromosomal walking by jumping large distances and to pinpoint the location of the DNA to be cloned much more precisely. No other model organism offered these powerful classical extensions of the standard molecular cloning repertoire.

Thus, it was the enormous experimental resources of pre-molecular *Drosophila* genetics that gave the fruit fly a head start over other experimental organisms like *C. elegans* or the mouse. Due to these resources, the chromosomal walking method was particularly powerful in the *Drosophila* system.[13] This work then provided an entry point to some genes and their products which, in all likelihood, could not have been isolated using other experimental organisms. The best example are the homeobox-containing genes. In order to demonstrate this, I will briefly review how these genes were discovered.

The homeobox was first identified as a sequence homology between the genes *Antennapedia*, *fushi tarazu*, and *Ultrabithorax* after these genes had been cloned. This discovery was made independently in two laboratories, namely by Michael Levine, Ernst Hafen and William McGinnis, who were working with Gehring in Basel, and Matthew Scott, who was working in the laboratories of Thomas Kaufman and Barry Polisky at Indiana. The Basel group then went on to search for homeobox-containing genes in other organisms. To their great surprise, they found them in a large variety of animals, including humans. Their search was greatly facilitated by the availability of the clones from *Drosophila* (Gehring 1998: 49), which could be used for colony hybridization screens on genomic DNA libraries prepared from any organism (so-called zooblots). Even though homeotic mutants exist in organisms other than *Drosophila*, none of them offered the same experimental resources for localizing and isolating genes. Thus, the genes isolated from *Drosophila* provided the research materials, in the form of DNA probes, necessary to isolate homologous homeobox genes in other organisms.

I suggest that the *production of research materials*—such as recombinant DNA clones—that can be used to search for homologous genes in other organisms is an as-yet underappreciated role of model organisms (see also Weber 2001). So far, most commentators have identified the role of such organisms as being either the constitution of research problems (the co-construction thesis, see Clarke and Fujimura 1992), or the extrapolation of theoretical knowledge to other organisms, especially humans (Schaffner 2001).

The cloning of any well-defined DNA fragment constituted a potential resource for further experimental work, because it could serve as a tool for cloning other DNA fragments. For example, a previously cloned DNA fragment could serve as a starting point of chromosomal walks to isolate sequences that are located in a nearby chromosomal region. In this way, cloned DNA fragments were used to clone other, totally unrelated sequences. In order to isolate homologous sequences, cloned DNA was used as a hybridization probe in colony hybridization screens. This method could be applied to DNA isolated from any organism, a fact that was crucial in isolating a host of genes from other organisms (including humans) using cloned DNA fragments from *Drosophila*. Thus, the more genes had been cloned, the easier it became to clone genes. With each new gene cloned, the *Drosophila* system was transformed from a breeder-reactor for isolating and mapping mutants into a molecular research tool for the production of specific recombinant DNA molecules.[14]

Of course, it must be emphasized that what I have described here is only a transient state in the development of *Drosophila* as an experimental system for molecular biology. It was the specific combination of traditional and new methods as well as concepts and research materials at a particular time and place that allowed geneticists to enter novel territory in developmental biology in the 1970s and early 1980s. Since then, the practice of molecular genetics has moved on, has been transformed by yet newer methods such as PCR (polymerase chain reaction) or automated DNA sequencing technology. To examine the role of these innovations is beyond the scope of this chapter. I expect that further study would reveal more examples of the typical interplay of historically grown methods, concepts, research materials and new technology that is characteristic of biological experimental systems (cf. Rheinberger 1997).

To conclude, classical mapping techniques played a crucial role in transforming the fruit fly into a powerful gene cloning tool. Using a notion introduced by Kohler (1991), we could describe *Drosophila* as a production system for specific DNA fragments that provided research materials for further study. Genetic and cytological maps, initially a convenient way of cataloging mutants and representing genetic structures, were assimilated into the preparative techniques of molecular biology.

In the final section, I will try to situate the cloning of the first fly genes in a different context, namely developmental biology.

Cloning and the quest for developmental genes

The adoption of recombinant DNA technology by *Drosophila* workers in the late 1970s and early 1980s had a lasting impact on developmental biology. As we have seen, combining these techniques with classical genetic and cytological mapping techniques gave geneticists access to genes and gene products such as the home-obox genes and the proteins they encode, which turned out to be of universal importance in all multicellular organisms. This finding was unexpected. Even though homeotic mutants had been known since the nineteenth century, it was widely assumed that this phenomenon had something to do with the segmented *Bauplan* of insects (Morange 2000: 196). The extremely strong evolutionary conservation of homeobox genes was a blessing for developmental biology, because it led to the discovery of developmental genes in organisms where they would otherwise have been much more difficult to find, such as the mammalian Hox genes (Gehring 1987). In this section, I show that the discovery of homeotic genes was a direct result of the cloning work described earlier.

The first homeotic mutant of *Drosophila* was described in 1915 by Calvin Bridges, who named it *Bithorax*. It transforms parts of the third thoracic segment into the anterior part of the second segment, thus mutants exhibit an extra pair of wings or part of a wing instead of the halteres. *Drosophila* geneticists did not immediately jump on the opportunities that homeotic mutations offered to study the genetic control of development. The usual explanation for this is that early geneticists suppressed questions of development in favor of chromosomal mechanics, because the latter were susceptible to a quantitative approach (Lawrence 1992: 211). Thus, homeotic mutants were not a subject for developmental studies until Edward Lewis at Caltech became interested in the *Bithorax* locus in 1946. Based on genetic analyses of a vast number of *Bithorax* mutants, he eventually developed a model of how the *Bithorax* locus controls segment identity in developing embryos (Lewis 1978).[15] Lewis's model involved eight genes, an estimate that turned out to be false. However, the idea that homeotic genes act combinatorially to determine segment identity in insects is still thought to be correct today.

The work of Lewis was essentially classical genetics and did not require any molecular techniques, nor any knowledge of the molecular mechanisms of inheritance. The same is true for the large-scale screen for mutations affecting early embryonic development in *Drosophila* carried out by Christiane Nüsslein-Volhard and Eric Wieschaus in the late 1970s (Keller 1996). However, it is clear that this work was important for establishing homeotic and other developmental genes as objects worthy for molecular studies. This is reflected by the fact that the homeotic gene complexes *Bithorax* and *Antennapedia* were among the first *Drosophila* genes to be cloned. At the time, the decision to clone a particular gene was a risky one, as the techniques were not quite as developed as they are today and the cloning and confirmation of a DNA fragment could take up several PhD theses worth of work.[16] The decision to invest time and funds into cloning a particular gene was a strategically important choice.

Not all genes were cloned for the same reasons. Some genes were cloned simply because there was an easy way of doing it, as was the case for the rRNA or heat shock genes. This is not to say that these genes are biologically uninteresting, but since all genes are to some extent biologically interesting and since research funding is a limited resource, there must be additional reasons why scientists choose to clone particular genes. Some genes were probably selected because they had been studied for a long time by classical geneticists and, therefore, there was a wealth of genetic data as well as well-defined mutant strains associated with them, for example *white* or *rudimentary*. In addition, some of these long-known genes have legendary status in the fly community, which may also have played a role.[17]

As far as the homeotic gene complexes *Bithorax* and *Antennapedia* are concerned, the laboratories of Hogness, Gehring and Kaufman spent considerable time and resources on their cloning—an investment that paid off. That the cloning of homeotic genes was high on the agenda of some *Drosophila* geneticists is evident in the case of Gehring, who mentioned this as a main reason for initiating cloning work in his laboratory (Gehring 1978). Gehring also recalls that, inspired by the famous work of F. Jacob and J. Monod on gene regulation in bacteria, he contemplated the isolation of *Drosophila*-analogs to DNA-binding proteins such as the *E. coli lac* repressor in the 1960s.[18] However, this seemed impossible then because of the extremely low concentrations of such proteins in *Drosophila* embryos. Later on, attempts were made in Gehring's laboratory to isolate specific DNA-binding proteins. However, the approach taken for isolating genes such as *Antennapedia* by chromosomal walking clearly proved to be most fruitful.

In Gehring's laboratory, the homeobox made its first appearance when clones from the *Antennapedia* region were hybridized to the entire collection of clones from the chromosomal walk (Gehring 1998: 47). Gehring points out that such cross-hybridizations were not entirely unexpected, as E.B. Lewis had argued for years that homeotic loci arose by gene duplications (see also Lawrence 1992: 216). The great surprise was that the hybridization signals observed were caused by an extremely conserved protein-coding region. Although there is some controversy surrounding the discovery of the homeobox, this is not relevant to the present analysis. What matters here is that the most important precondition for this discovery was the positional cloning of the homeotic gene complexes *Antennapedia* and *Bithorax*.

What is the relationship between the earlier genetic studies of development in *Drosophila* and the molecular work that led to the discovery of homeobox genes? I suggest that the early work of Lewis and others provided not just theoretical knowledge concerning the function of homeotic gene complexes. The main advantage that *Drosophila* offered for isolating developmental genes lay in the enormous *experimental resources* that came with this organism. These resources include, first, highly detailed genetic and cytological maps and the associated techniques, that made localizing developmental gene complexes a feasible task. By combining these mapping techniques with recombinant DNA technology, it was possible to isolate and identify DNA fragments harboring these gene complexes. Second, *Drosophila* workers had isolated thousands of mutant strains that were used by molecular biologists for chromosomal walking and jumping, and for pinpointing cloned DNA fragments more precisely. As we have, seen Hogness's group used *Bithorax* mutants isolated by Edward Lewis to clone this gene complex. Gehring's laboratory also used *Drosophila* mutants that had been isolated and characterized by classical techniques—such as the inversion mutant *Humeral*—in order to clone genes from the homeotic *Antennapedia* complex. The homeobox was bound to be discovered once cloned DNA fragments harboring developmental genes such as *Ultrabithorax*, *Antennapedia*, and *fushi tarazu* became available.

To sum up, *Drosophila* played such an important role in the molecularization of developmental biology not so much because of antecedently existing theoretical knowledge on its embryology, but mainly due to the enormous experimental resources accumulated by generations of geneticists. Classical mapping techniques and detailed chromosome maps, as well as *Drosophila* mutants isolated and characterized by classical methods, were an essential part of these experimental resources. By combining them with recombinant DNA technology, *Drosophila* workers were able to mobilize these resources and to transform the fruit fly into a powerful tool for cloning developmental genes.

Acknowledgments

I wish to thank Walter Gehring for granting me an interview and for providing the illustrations; Hans-Jörg Rheinberger and Jean-Paul Gaudillière for organizing a series of great workshops on the history and philosophy of genetics, from which the present work has benefited greatly; and Angela Creager, Jay Aronson, Paul Hoyningen-Huene and Daniel Sirtes for helpful comments. This chapter has also benefited from discussions in Kenneth Waters's graduate student seminar in philosophy of biology at the University of Minnesota (Fall 2000).

Notes

1 See for example Ankeny (2000), Burian (1992, 1993, 1996), de Chadarevian (1998), Clause (1993), Creager (2002), Geison and Creager (1999), Holmes (1993), Kohler (1991, 1993), Lederman and Burian (1993), Lederman and Tolin (1993), Mitman and Fausto-Sterling (1992), Rader (1999), Rheinberger (1997, 2000), Schaffner (2001), Summers (1993), Weber (2001, in press), Zallen (1993)
2 The term molecular cloning is also used. It has the advantage of avoiding confusion with Dolly the sheep.
3 Ribosomes are protein-synthesizing particles, which consist of protein and RNA. These RNA molecules are differentiated by their size, so 18S and 28S are rRNA molecules of different size. rRNA is encoded by special genes. The work described here allowed molecular biologists to isolate these genes.
4 In molecular biology, hybridization refers to the experimental production of double-stranded nucleic acid molecules where the two strands are derived from different sources. The technique makes use of the strong disposition of complementary nucleic acid molecules to anneal to a double helix (which is also crucial for their informational properties). Under certain conditions, it is possible to hybridize single-stranded RNA to

a complementary DNA molecule, which forms the basis of colony hybridization and many other molecular techniques.

5 Gehring contracted the cloning know-how in the form of a post-doctoral researcher, Paul Schedl, who had been trained at Stanford (Gehring 1998: 41).

6 mRNA is short for messenger-RNA and designates the RNA copies made from the genes in the cell nucleus. The function of mRNA is to serve as a template in the synthesis of proteins (the sequence of nucleotide bases in RNA determines the sequence of amino acids in a growing protein molecule). The concentration of different mRNAs can vary greatly inside a cell, depending on the cell's metabolic needs. This effect was exploited in order to isolate the mRNA for the heat shock proteins.

7 This is another important technique that uses the unique distribution of cleavage sites for different restriction enzymes in order to physically map DNA molecules. The DNA fragment to be mapped is cut with a number of restriction enzymes and the resulting fragments are separated by molecular weight by gel electrophoresis. Since the molecular weight of the fragments is proportional to their length, the specific pattern of restriction fragments can be used to construct a map of the restriction sites on the DNA fragment.

8 Actually, the products of these genes were known from classical studies in biochemical genetics (xanthine dehydrogenase and acetylcholinesterase, respectively). However, this knowledge could not then be used for cloning because the concentrations of the corresponding mRNAs were too low to permit isolation of specific RNAs for colony hybridization. By contrast, some other household genes like *Adh* (alcohol dehydrogenase) were cloned by prior biochemical identification of an mRNA species (e.g. Benyajati *et al.* 1980).

9 This appears to have been a competitive rather than collaborative effort. The *Drosophila* community at that time seems to have been different from the *C. elegans* community, which exhibited a strong culture of cooperativeness (de Chadarevian 1998 and Chapter 5, this volume). The reasons for these cultural differences are unknown to me.

10 The laboratory of Thomas Kaufman at Indiana University used a somewhat different strategy, which also involved much chromosomal walking, to isolate clones spanning the *Antp* region (Scott *et al.* 1983).

11 cDNA stands for complementary DNA. It is experimentally produced by treating an RNA template with the enzyme reverse transcriptase, which synthesizes DNA molecules which are complementary to the RNA template.

12 Another technique called transposon-tagging exploited the presence, in some genes, of mobile genetic elements (transposons) for cloning. For example, the gene *white* was cloned by this method, which was one of the first genes to be discovered by T.H. Morgan in 1910.

13 Positional cloning was also possible in other organisms (as well as in cloning human genes), but the *Drosophila* work appears to have taken the lead, at least in the early 1980s.

14 This production aspect of some research in molecular biology has been noted by Robert Kohler (1991) and Karin Knorr-Cetina (1999: esp. ch. 6). These authors seem to take the industrial–technological metaphors of "production," "instruments," "tools" and "machines" quite literally. I think that these metaphors (even if harmless) are not very helpful, therefore, I propose to talk about *preparative experimentation* (Weber, in press, ch. 6).

15 This work earned Lewis a Nobel Prize, which he shared with Christiane Nüsslein-Volhard and Eric Wieschaus in 1995.

16 Initially, the successful molecular cloning of a gene by a research group merited publication in a first-rate journal. Later, as the techniques became more developed, such journals began to require that some biologically interesting results be included in a paper reporting the cloning of a gene.

17 W. Gehring (interview, August 19, 2000). When I asked him why he and his co-workers chose the *white* locus for one of their first cloning attempts, Gehring half-jokingly

replied that this gene had a "sacred meaning" to him, because it was one of the first genes to be described by T.H. Morgan. A large portrait of Morgan hangs over the door in Gehring's office. This episode suggests that tradition is more important in science than is generally appreciated.

18 Interview, August 19, 2000.

Bibliography

Adams, M.D. *et al.* (2000) "The Genome Sequence of *Drosophila melanogaster*," *Science*, 287: 2185–95.

Ankeny, R.A. (2000) "Fashioning Descriptive Models in Biology: Of Worms and Wiring Diagrams," *Philosophy of Science* (Proceedings), 67: S260–S272.

Artavanis-Tsakonas, S., Schedl, P., Tschudi, C., Pirrotta, V., Steward, R., and Gehring, W.J. (1977) "The 5S *Genes of Drosophila melanogaster*," *Cell*, 12: 1057–67.

Artavanis-Tsakonas, S., Schedl, P., Mirault, M.-E., Moran, L., and Lis, J. (1979) "Genes for the 70,000 Dalton Heat Shock Protein in Two Cloned *D. melanogaster* DNA Segments," *Cell*, 17: 9–18.

Bender, W., Spierer, P., and Hogness, D.S. (1983) "Chromosomal Walking and Jumping to Isolate DNA from the Ace and Rosy Loci and the Bithorax Complex in *Drosophila melanogaster*," *Journal of Molecular Biology*, 168: 17–33.

Benyajati, C., Wang, N., Reddy, A., Weinberg, E., and Sofer, W. (1980) "Alcohol Dehydrogenase in Drosophila: Isolation and Characterization of Messenger RNA and cDNA Clones," *Nucleic Acids Research*, 8: 5649–67.

Burian, R.M. (1992) "How the Choice of Experimental Organism Matters: Biological Practices and Discipline Boundaries," *Synthese*, 92: 151–66.

—— (1993) "How the Choice of Experimental Organism Matters: Epistemological Reflections on an Aspect of Biological Practice," *Journal of the History of Biology*, 26: 351–67.

—— (1996) " 'The Tools of the Discipline: Biochemists and Molecular Biologists': A Comment," *Journal of the History of Biology*, 29: 451–62.

Chadarevian, S. de (1998) "Of Worms and Programmes: *Caenorhabditis elegans* and the Study of Development," *Studies in History and Philosophy of Biological and Biomedical Sciences*, 29C: 81–106.

Clarke, A.E. and Fujimura, J.H. (1992) "What Tools? Which Jobs? Why Right?" in A.E. Clark and J.H. Fujimura (eds) *The Right Tools for the Job. At Work in Twentieth-Century Life Science*, Princeton, NJ: Princeton University Press.

Clause, B.T. (1993) "The Wistar Rat as a Right Choice: Establishing Mammalian Standards and the Ideal of a Standardized Mammal," *Journal of the History of Biology*, 26: 329–49.

Creager, A.N.H. (2002) *The Life of a Virus: Tabacco Mosaic Virus as an Experimental Model, 1930–1965*, Chicago, IL: University of Chicago Press.

Falk, R. (2004) "Applying and Extending the Notion of Genetic Linkage: The First Fifty Years," in H.-J. Rheinberger and J.-P. Gaudillière (eds) *Classical Genetic Research and its Legacy: The Mapping Cultures of Twentieth-century Genetics*, London: Routledge.

Gannett, L. and Griesemer, J.R. (2004) "Classical Genetics and the Geography of Genes," in H.-J. Rheinberger and J.-P. Gaudillière (eds) *Classical Genetic Research and its Legacy: The Mapping Cultures of Twentieth-century Genetics*, London: Routledge.

Garber, R.L., Kuroiwa, A. and Gehring, W.J. (1983) "Genomic and cDNA Clones of the Homeotic Locus Antennapedia in Drosophila," *The EMBO Journal*, 2: 2027–36.

Gehring, W.J. (1978) "Establishment of a Drosophila Gene Bank in Bacterial Plasmids: Elizabeth Goldschmidt Memorial Lecture," *Israel Journal of Medical Sciences*, 14: 295–304.

—— (1985) "The Molecular Basis of Development," *Scientific American*, 253: 152–62.

—— (1987) "Homeo Boxes in the Study of Development," *Science*, 236: 1245.

—— (1998) *Master Control Genes in Development and Evolution: The Homeobox Story*, New Haven, CT: Yale University Press.

Geison, G.L. and Creager, A.N.H. (1999) "Introduction: Research Materials and Model Organisms in the Biological and Biomedical Sciences," *Studies in History and Philosophy of Biological and Biomedical Sciences*, 30C: 315–18.

Grunstein, M. and Hogness, D.S. (1975) "Colony Hybridization: A Method for the Isolation of Cloned DNAs That Contain a Specific Gene," *Proceedings of the National Academy of Sciences of the United States of America*, 72: 3961–5.

Holmes, F.L. (1993) "The Old Martyr of Science: The Frog in Experimental Physiology," *Journal of the History of Biology*, 26: 311–28.

Keller, E.F. (1996) "Drosophila Embryos as Transitional Objects: The Work of Donald Poulson and Christiane Nüsslein-Volhard," *Historical Studies in the Physical and Biological Sciences*, 26: 313–46.

Knorr-Cetina, K. (1999) *Epistemic Cultures: How the Sciences Make Knowledge*, Cambridge, MA: Harvard University Press.

Kohler, R.E. (1991) "Systems of Production: Drosophila, Neurospora and Biochemical Genetics," *Historical Studies in the Physical and Biological Sciences*, 22: 87–129.

—— (1993) "Drosophila: A Life in the Laboratory," *Journal of the History of Biology*, 26: 281–310.

—— (1994) *Lords of the Fly. Drosophila Genetics and the Experimental Life*, Chicago, IL: University of Chicago Press.

Kornberg, T.B. and Krasnow, M.A. (2000) "The Drosophila Genome Sequence: Implications for Biology and Medicine," *Science*, 287: 2218–20.

Lawrence, P.A. (1992) *The Making of a Fly: The Genetics of Animal Design*, London: Blackwell Scientific.

Lederman, M. and Burian, R.M. (1993) "Introduction: The Right Organism for the Job," *Journal of the History of Biology*, 26: 235–7.

Lederman, M. and Tolin, S.A. (1993) "OVATOOMB: Other Viruses and the Origins of Molecular Biology," *Journal of the History of Biology*, 26: 239–54.

Lewis, E.B. (1978) "A Gene Complex Controlling Segmentation in Drosophila," *Nature*, 276: 565–70.

Mitman, G. and Fausto-Sterling, A. (1992) "Whatever Happened to Planaria? C.M. Child and the Physiology of Inheritance," in A.E. Clarke and J.H. Fujimura (eds) *The Right Tools for the Job. At Work in Twentieth-Century Life Sciences*, Princeton, NJ: Princeton University Press.

Moran, L., Mirault, M.E., Tissieres, A., Lis, J., Schedl, P., Artavanis-Tsakonas, S., and Gehring, W.J. (1979) "Physical Map of Two *D. melanogaster* DNA Segments Containing Sequences Coding for the 70,000 Dalton Heat Shock Protein," *Cell*, 17: 1–8.

Morange, M. (1998) *A History of Molecular Biology*, Cambridge, MA: Harvard University Press.

—— (2000) "The Developmental Gene Concept: History and Limits," in P. Beurton, R. Falk, and H.-J. Rheinberger (eds) *The Concept of the Gene in Development and Evolution*, Cambridge: Cambridge University Press.

Rader, K.A. (1999) "Of Mice, Medicine, and Genetics: C.C. Little's Creation of the Inbred Laboratory Mouse," *Studies in History and Philosophy of Biological and Biomedical Science*, 30C: 319–44.

Rheinberger, H.-J. (1997) *Toward a History of Epistemic Things: Synthesizing Proteins in the Test Tube*, Stanford, CT: Stanford University Press.

—— (2000) "Ephestia: The Experimental Design of Alfred Kühn's Physiological Developmental Genetics," *Journal of the History of Biology*, 33: 535–76.

Rubin, G.M. and Lewis, E.B. (2000) "A Brief History of Drosophila's Contributions to Genome Research," *Science*, 287: 2216–18.

Schaffner, K.F. (2001) "Extrapolation from Animal Models. Social Life, Sex, and Super Models," in P.K. Machamer, R. Grush, and P. McLaughlin (eds) *Theory and Method in the Neurosciences*. Pittsburgh-Konstanz Series in the Philosophy and History of Science, Pittsburgh, CA: University of Pittsburgh Press.

Schedl, P., Artavanis-Tsakonas, S., Steward, R., Gehring, W.J., Mirault, M.E., Goldschmidt-Clermont, M., Moran, L., and Tissieres, A. (1978) "Two Hybrid Plasmids with *D. melanogaster* DNA Sequences Complementary to mRNA Coding for the Major Heat Shock Protein," *Cell*, 14: 921–9.

Scott, M.P., Weiner, A.J., Hazelrigg, T.I., Polisky, B.A., Pirrotta, V., Scalenghe, F., and Kaufman, T.C. (1983) "The Molecular Organization of the Antennapedia Locus of Drosophila," *Cell*, 35: 763–76.

Summers, W.C. (1993) "How Bacteriophage Came to Be Used by the Phage Group," *Journal of the History of Biology*, 26: 255–67.

Weber, M. (1998) "Representing Genes: Classical Mapping Techniques and the Growth of Genetic Knowledge," *Studies in History and Philosophy of Biological and Biomedical Sciences*, 29: 295–315.

—— (2001) "Under the Lamppost: Commentary on Schaffner," in P.K. Machamer, R. Grush, and P. McLaughlin (eds) *Theory and Method in the Neurosciences*, Pittsburgh-Konstanz Series in the Philosophy and History of Science, Pittsburgh: University of Pittsburgh Press.

—— (in press) *Philosophy of Experimental Biology*. Cambridge: Cambridge University Press.

Zallen, D.T. (1993) "The 'Light' Organism for the Job: Green Algae and Photosynthesis Research," *Journal of the History of Biology*, 26: 269–79.

4 Gene expression maps

The cartography of transcription

Scott F. Gilbert

Mapping communities

"Mapping communities in genetics" can mean different things depending upon which branch of genetics one apprehends. Mapping is a metaphor referring to the practice of cartography, the construction of a two-dimensional representation of a three-dimensional surface. In their public uses, maps attempt to accurately scale the political or geographic dimensions of a territory so that the possessor of the map can obtain an overview of the political or physical landscape. Maps emphasize boundaries and distinctions—where does the land terminate and the sea begin, at what point do the laws of Germany end and the laws of France take effect? Needless to say, map making has always been a political act and no such representation is ever completely without its abstractions and exaggerations (Monmonier 1991).

So what has mapping to do with genetics? In population genetics, gene mapping refers to allele frequency distribution over space. Thus, Cavelli-Sforza and colleagues call their book *History and Geography of Human Genes*. Lisa Gannett and Jim Griesemer discuss this tradition in volume 1 (Chapter 4) of this publication. To a developmental geneticist, mapping refers to representations of gene expression over the space of an embryo or part of an embryo. Thus, developmental genetics has a major program in mapping gene expression. Third, in medical and transmission genetics, maps are used to indicate the relative positions of genes on a chromosome. "Genomic geography" is the keyword for chromosome linkage "maps" at a US Department of Energy website called "Know Ourselves." This site also talks about "Exploring the Genomic Landscape," and it informs us that "one of the central goals of the Human Genome Project is to produce a detailed 'map' of the human genome. But, just as there are topographic maps and political maps and highway maps of the United States, so there are different kinds of genome maps, the variety of which is suggested in Genomic geography." It is interesting that this site problematizes the map metaphor with its quotation marks.[1]

As a developmental geneticist, the program to make unidimensional "maps" of chromosomes seems more akin to census taking than map making. "The gene for cystic fibrosis resides at chromosome 7q31.2." "The actin gene in *Drosophila* can be found at position 88F4–5." These are addresses along single streets. Census taking is dull, but mapping is romantic exploration. This can be seen in the advertisements that extrapolate from the mapping metaphor. In the advertisement for DoubleTwist (2000) caravels sail on the undulating waves of the double helix, where one must "navigate vast seas of data" and where there is "no better compass for navigating the genomics frontier than DoubleTwist.com." The advertisement for New England BioLabs (2001) shows an antiquarian map of Africa superimposed upon the body of a woman. The connotation is that in knowing the genome, we will know the phenotype—the human woman/Africa/ the natural. Is the genotype map like the secret rutters of the Portuguese and Spanish explorers: Private knowledge that yields enormous wealth for both company and country? Do we still believe that the genome is the blueprint of the phenotype? The romantic conception of discovery, wealth, and the old deterministic aspect of genetics hide uneasily in the mapping metaphor.

Fate maps: placing the future onto the past

Whatever the appropriateness or inappropriateness of the map metaphor to chromosomal genetics, this mapping metaphor has an honored history in developmental genetics. Mapping programs are and have been a standard part of embryology and contemporary developmental biology. The "fate map" is the representation of future fate superimposed onto the structure of an earlier stage embryo. In other words, the embryologist makes a representation of *what is to be* and superimposes these data *onto a surface of what is yet to develop*. For instance, fate maps such as those of the tunicate egg in Figure 4.1 were made by E.G. Conklin. In 1905, Conklin reported the lineages of the tunicate *Styela partita*, being able to trace each cell until it only produced one type of progeny (Figure 1(a)). This can only be performed on embryos having a small number of cells and which differentiate very quickly, and *Styela* was extremely useful in that its blastomeres differed by size and color. These lineages were confirmed by others, and a map was made in the 1930s (Figure 1(b)) wherein these lineages were superimposed around the egg and early embryo.

In vertebrate embryos, scientists made partial fate maps that dealt with entire areas rather than individual cells. The fate map in Figure 4.2 shows the locations of various morphogenetic fields in the salamander neurula. Here, these fields were defined by the ability of those cells to produce the fated structure when transplanted into another region of the embryo, the absence of those structures when such cells were removed, and the ability of cells within the field to regulate (Huxley and De Beer 1934).

In 1929, Walther Vogt constructed the first systematic fate map of a developing vertebrate, when he applied vital dyes to the early amphibian embryo. He impregnated agar chips with a vital dye and then placed the early embryo into a wax depression such that the chips would impart some of their dye onto the embryo. In this way, he was able to mark cells early in development and see what structures these cells would later become. He then demarcated the fate of the regions upon the early embryo (blastula) such that one could see what the cells at any given location would become (Figure 4.3). It is easy to see why these reconstructions were

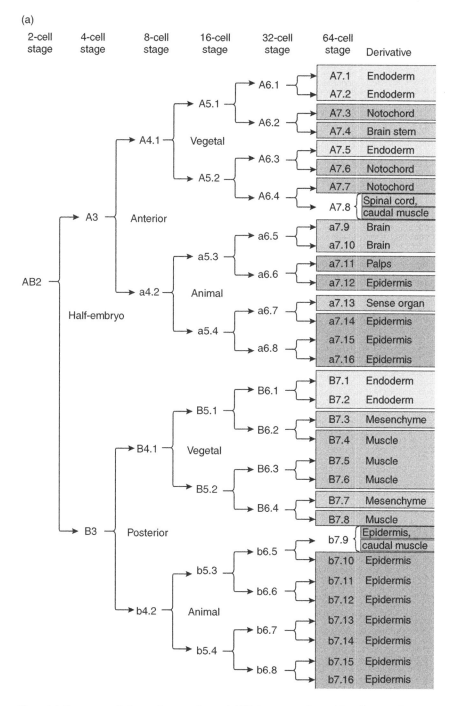

Figure 4.1 Fate map of the tunicate embryo. (a) Dichotomous branching lineage diagram of the tunicate *Styela*. (b) Fate map constructed by superimposing the fates onto the one-cell and eight-cell embryo. These and the other illustrations are taken from Gilbert (2000) to avoid copyright problems.

(b)

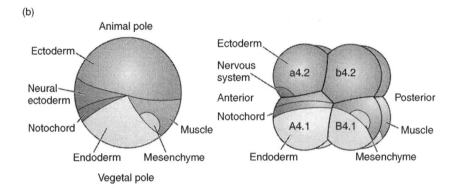

Figure 4.1. Continued.

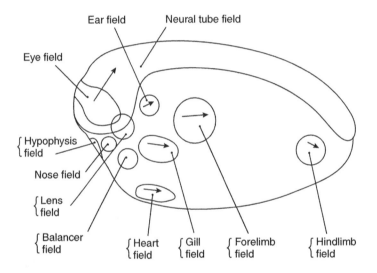

Figure 4.2 Fate map of an amphibian neurula, showing morphogenetic fields that will, as the salamander develops, produce the various larval structures. The arrows within the fields indicate that they are polarized (After Huxley and De Beer 1934).

called fate maps. Due to the spherical nature of the amphibian embryo, Vogt's fate maps looked very much like globes. They had two poles—the animal pole and the vegetal pole—and an "equator" along the center, which served as a landmark. The eggs divide orthogonally, marking longitudinal and latitudinal positions with respect to the point of sperm entry (future ventral surface) and a conveniently colored cytoplasm, the grey crescent (the future dorsal surface). There is even a prime meridian, the cleavage furrow that forms first and which divides the egg into its future left and right sides.

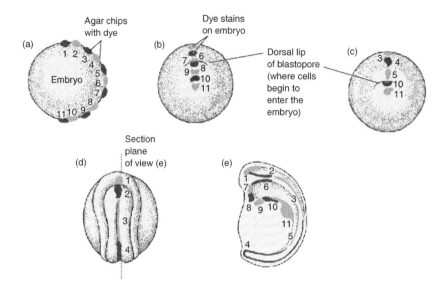

Figure 4.3 Vital staining of newt embryo to produce a fate map. (a) Vogt's method for marking specific regions of the embryo with dye-impregnated agar chips. (b–d) Dorsal surface views of successively later embryos. (e) Newt embryo dissected to show stained cells in the interior.

Good maps, whether political or embryological, are critical for making discoveries. As Viktor Hamburger (1988) pointed out, Hans Spemann mis-interpreted the results of his first dorsal blastopore lip transplantations, since he used an old fate map of the newt blastula. Because of this relatively poor map, he had thought that the dorsal blastopore lip cells differentiated into ectoderm. Thus, there was nothing very special about this structure. However, when Peterson told Spemann that Spemann's own older research had concluded that the dorsal blastopore lip formed mesoderm, the observation of new axis formation became much more interesting. The confirmation of that later interpretation—that the dorsal blastopore lip involuted into the embryo, became notochordal tissue, and induced the dorsal ectoderm to become the neural plate (Spemann and Mangold 1924)—became one of the most important experiments in modern embryology.

Interestingly, the geomorphic amphibian fate map of Vogt turned out not to be universally applicable to all amphibians. One of the cartographic conventions of such maps gave rise to the erroneous view that the deep cells had the same fate as the cells of the surface. This was found to be false in *Xenopus*. Here, the fate map of the *internal* cells differed from the fate map of the *external* cells (Keller 1976). Fate maps are being refined and revised for well-studied organisms, and they are being constructed anew for organisms that are just beginning to be studied. Fate maps are still giving us new information about developmental boundaries (see Sherwood and McClay 2001), are still topics of argument (see Bauer *et al.* 1994; Lane and Smith 1999), and are now being used to provide new knowledge about evolution.

When fate maps are compared, evolutionary change can be inferred. For instance, Greg Wray and Rudy Raff (1990) compared fate maps to show the changes in blastomere fate during early development of direct developing sea urchins. They showed that the fate map of the direct developing urchins had been changed such that the cells that would have formed larval ectoderm are now forming the vestibular structures that become the adult. Moreover, the positions of neuronal precursors has been altered. The fact that these species diverged only around 10 million years ago showed that significant developmental change can occur by altering early development, something that had formerly been thought to be highly improbable. The fate-mapping program of developmental biology is very much alive and well.

Gene expression maps

A related, albeit chronologically later, program has been the mapping of gene expression patterns. There are many ways to map gene expression, and they usu-ally give a direct projection upon an embryonic surface. The most common way is through *in situ hybridization*. Here, one can actually stain for the accumulation of a particular type of mRNA. This ability has revolutionized developmental biol-ogy, bringing not only cell-level resolution to the analysis of transcription pat-terns, but also mandating that the scientist have knowledge of embryological anatomy. The probe and its reporter accumulate only at the sites where the target mRNA has accumulated and thus can be found. A second way to map gene transcription patterns is to use a reporter gene sequence (one that can be readily identified and which is not usually made in the animal) and connect it to the regulatory region of a particular gene. Wherever the gene is usually expressed, the reporter gene will be transcribed.

One of the consequences of these gene expression maps was to show that embryonic development did not necessarily use the same building units as one sees in adult anatomy and physiology. Thus, gene expression patterns were seen in specific domains—*Drosophila* compartments and parasegments; mammalian rhombomeres and enamel knots—that have no anatomical correlates in the adult. Without these maps of gene expression patterns, we probably would not be aware of these entities.

Gene expression maps can serve several functions, one of the most obvious is to use them as one would use fate maps. For instance, if one knows that the noto-chordal cells of *Xenopus* transcribe the *chordin* gene, a probe to *Xenopus chordin* can be used to locate the cells that will produce the notochord. Indeed, it was exactly this type of gene expression mapping that gave developmental biologists both a fine-resolution map of the amphibian organizer and, at the same time, provided a mechanism for the initiating movements of gastrulation. In *Xenopus*, the cells of the organizer ultimately contribute to four cell types—pharyngeal endoderm, head mesoderm (prechordal plate), dorsal mesoderm (primarily the notochord), and the chordaneural hinge (Keller 1976; Gont *et al.* 1993). The pharyngeal endoderm and prechordal plate lead the migration of the organizer tissue and appear to induce the forebrain and midbrain. The dorsal mesoderm induces the hindbrain and spinal cord, and the chordaneural hinge induces the tip of the tail. The expression patterns of transcription factors provided the key to how this was done. Vodicka and Gerhart (1995) pioneered the study of cell-specific gene expression in the organizer region, and this was extended by the work of Winklbauer and Schürfeld (1999) (Figure 4.4). The latter authors noted that the *Cerberus*-expressing cells of the pharyngeal endoderm were lying on the blastocoel floor and were not in any position to lead the *goosecoid*-expressing head mesoderm and the *cordin*-expressing notochordal cells along the dorsal surface of the embryo. So they predicted and observed the rotation of the yolky vegetal hemisphere cells, which was required to put the *Cerberus*-expressing pharyngeal endoderm cells dorsal and anterior to those cells expressing *chordin* or *goosecoid*.

Gene expression maps need not be confined to fate maps, however. They can be used to show signaling. One of the most important uses of gene expression

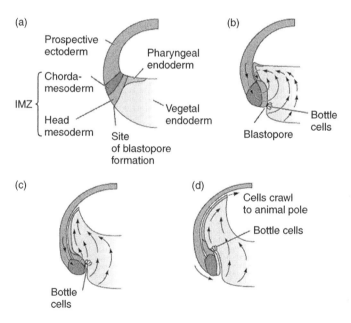

Figure 4.4 Early movements of *Xenopus* gastrulation, as shown by gene expression. The cells fated to form the pharyngeal endoderm express *cerberus*. These cells are at the anteriormost point of the migrating epithelium, and they are moved into this position by the rotation of the deeper vegetal cells. The cells fated to become head mesoderm express the *goosecoid* gene, and they follow the pharyngeal endoderm. The cells fated to become chordamesoderm are expressing *Xbra* (*Xenopus brachyury*) and they follow the other two regions.

mapping has been to show which cells are signaling centers. In this analysis, the gene being expressed encodes a paracrine factor that is capable of influencing the development of other cells. For instance, a small block of mesodermal cells at the posterior junction of the limb bud and the body wall has the ability to polarize the chick limb along the anterior–posterior axis. When such tissue is taken from the posterior margin of one limb bud and placed into the anterior margin of a second limb bud, the host limb bud develops two mirror-image sets of digits. This region was called the zone of polarizing activity (ZPA). In 1993, Riddle and his colleagues in Cliff Tabin's laboratory showed that ZPA was defined by the expression of the *Sonic hedgehog* gene. First, they showed that the sonic hedgehog protein is necessary and sufficient to account for the ZPA's activities. When they caused this protein to be synthesized in the anterior margin of a limb bud, they obtained mirror image duplications. More interestingly, they were able to correlate the time, place, and amount of *Sonic hedgehog* gene expression with the classically defined potency of this region to induce the mirror-image duplications.

The use of gene expression maps in evolutionary developmental biology

The temporal priority of the gene expression map over the fate map is seen when genes are functionally deleted from embryos. In these cases, the fate map changes as well (Kishimoto *et al.* 1997) (Figure 4.5). The cell-lineage map, the fate map, and the gene expression map have been united in this type of experiment, with the gene expression map having priority and explaining the others.

This relationship between the gene expression map and the fate map is critical for evolutionary developmental biology. One of the tenets of "evo-devo" has been that evolution is predicated upon hereditable changes in development. These changes are therefore seen as changes in gene expression. So gene expression

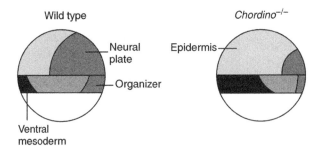

Figure 4.5 Change in fate map of zebrafish when the *chordino* gene is deleted. The chordino product is important in forming neural (dorsal) ectoderm and dorsal mesoderm. In the absence of this gene, the epidermis expands at the expense of the neural plate and the ventral and lateral mesoderm expand at the expense of the dorsal (organizer) mesoderm.

changes and continuities are the stuff of what much of evo-devo has been made. The case for the inclusion of evolutionary developmental biology into evolutionary biology has been made largely upon the changes and the continuities of gene expression maps. Thus, comparative gene expression maps have played key roles as evidence for the importance of evolutionary developmental biology.

One of the first of these comparative gene expression maps showed the homologous expression of the *Hox* genes and head transcription factors (see Hirth and Reichert 1999) (Figure 4.6) between the protostomes (represented by the arthropod *Drosophila)* and the deuterostomes (represented by the vertebrates Gallus and *Mus).* This research stressed the similarities of the protostomes and the deuterostomes, showing that not only did the two groups have homologous *Hox* genes, not only were these genes on the same order in their respective chromosomes, but the geographic order of each gene's expression was also conserved. The expression pattern of *Hox* genes throughout the animal kingdom was so similar that Slack and colleagues (1993) proposed that this constituted the fundamental basis of being an animal. Interestingly, more recent gene expression maps—those presenting *Hox* gene expression in sponges (Schierwater and Kuhn 1998)—have been used as evidence against this view.

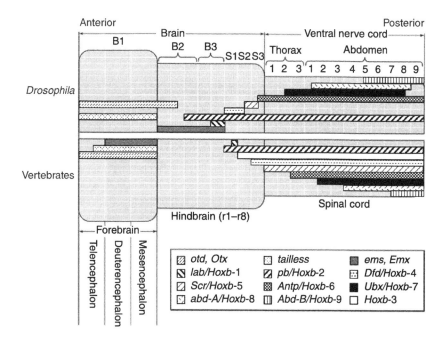

Figure 4.6 Expression of regulatory transcription factors in *Drosophila* and vertebrates. The *Drosophila* genes *ems, tll,* and *otd* are homologous to the vertebrate *Emx, Tll,* and *Otx* genes, respectively. The *Hox* genes expressed in *Drosophila* and in vertebrates have similar patterns in their respective hindbrains and neural cords.

Moreover, further investigations demonstrated that variations of this "standard" expression pattern could produce morphological changes. Changes in the *Hox* gene expression map in crustaceans were correlated with the changing number of maxillipeds, and changes in the *Hox* gene expression pattern in vertebrates correlated with changes in the number of cervical vertebrae (Gaunt 1994; Burke *et al.* 1995; Averof and Patel 1997). In some cases, severe alterations all but eliminated the certain constellations of *Hox* gene expression, and these eliminated certain types of vertebrae. Thus, in snakes, the *Hox* gene expression pattern allows the thoracic (ribbed) vertebrae to expand at the expense of the cervical (neck) and lumbar (abdominal) vertebral types. This eliminates the boundary conditions necessary for forelimb formation (Cohn and Tickle 1999).

Alterations of the *Hoxd11* and *Hoxd13* expression pattern have even been proposed to account for the formation of the vertebrate autopod (wrist/ankle and digit bones). Figure 4.7 depicts a Devonian transition of about 350 million years ago. Although there is no way of knowing the *Hox* gene expression of the lobe-finned ancestor fish (or how it might be fruitfully compared to a highly derived modern osteichthyan such as *Danio rerio*), the alterations in gene expression pattern between modern fish and modern tetrapods and the importance of *Hox* genes in specifying limb parts converged to make the different gene expression maps important evidence for a mechanism by which fish fins could be transformed into tetrapod limbs (Sordino *et al.* 1995; Shubin *et al.* 1997).

Gene expression maps can also provide clues as to how genes used in one area of development can be recruited to another. For instance, the *Distal-less* gene, used

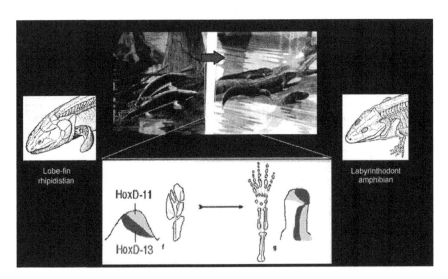

Figure 4.7 In the Devonian, labyrinthodont amphibian tetrapods emerged from an ancestor that was probably similar to lobe-fin rhystidipians. The change in *Hox* gene expression patterns at the end of the fin have been proposed to have created the autopod and enabled limbs to arise from the stalk of fins.

to define the limb primordia of insects became co-opted to produce the eyespots of butterfly wings (Brakefield *et al.* 1996; Beldade *et al.* 2002), and the *fgf10* gene, used in producing the tetrapod limb bud may also be utilized for producing the carapacial ridge specific to turtles (Loredo *et al.* 2001).

Gene expression maps can also be used to suggest homologous relationships that are not obvious. For instance, regulatory regions from *amphioxus Hox* genes expressed in the anterior neural tube were found to drive the spatially localized expression of a reporter gene in vertebrate neural crest cells and anterior neurogenic placodes despite the fact that these cephalochordates lack both the neural crest and the anterior placodes (Manzanares *et al.* 2000). This implies that there was already the expression pattern of "head" before there were cephalic structures, and that these anterior regions of the cephalopod neural tube are therefore homologous to vertebrate brains. Similar research has also homologized the amphioxus endostyle to the vertebrate thyroid gland. In a recent paper, Meinhardt (2002) has used gene expression data to propose the hypothesis that the entire *body* of the diploblastic and radially symmetrical *Hydra* is homologous to the *brain* of triploblastic and bilaterally symmetric animals. Moreover (counter to our intuition that the tentacled region makes the "head" of the hydra), the most anterior regions of animal heads would have been formed from the *pedal* (*Nk2.1*-expressing) domains of a *Hydra*-like ancestor. Comparative gene expression mapping forms the evidence for his entire thesis.

New techniques of gene expression mapping

Computer-aided technology has become increasingly important in the gene mapping community. Some of the greatest advances are in the area of three-dimensional expression mapping. The first advance concerns three-dimensional reconstructions wherein the computer rapidly and objectively realigns embryonic sections so as to make them usable for three-dimensional reconstruction (Figure 4.8)

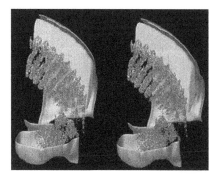

Figure 4.8 Fate mapping of myotome cells (dark gray) as they enter the limb bud. Computer reconstruction made from *myf-5* expression in the developing myotome cells (Courtesy: J. Streicher and G. Müller).

(Streicher *et al.* 2000). This procedure combines the methodological advantages of whole mount *in situ* hybridization and the high scale resolution of serial sectioning. After capturing phase-contrast and brightfield views of serial sections, it can map *in situ* hybridization data through a series of algorithms. The "subjective" interactions of the processor have been eliminated.

The internet has also become an important tool in gene expression mapping. For instance, the Kidney Development Database (http://golgi.ana.ed.ac.uk/kidhome.html), coordinated by Jonathan Bard and James Davies, allows one to search not only by gene but also by the part of the developing kidney. Accessing the "3D Confocal Reconstruction of Gene Expression in Mouse (Hecksher-Sørensen and Sharpe 2002) allows one to rotate embryos to show the pattern of their gene expression. Movies can be made of changes in gene expression patterns, and one can log on and access such phenomena as mouse kidney branching, visualized by placing the green fluorescent protein reporter gene onto the *Hoxb7* promoter (Srinivas *et al.* 1999; Constantini 2002).

Another four-dimensional aid is the use of Geographic Imaging System (GIS) analysis. Epoxy resin casts of an organ (such as a tooth) are optically sectioned using a laser confocal microscope. High-resolution digital elevation models (DEMs) of the organ topology are produced from the image stacks using the 3Dview version of the NIH-Image software. Digital elevation models can be transferred to GIS software as well as interpreted by surface rendering computer programs. All traditional morphometrical measurements can be obtained from DEMs. However, the total shape data (i.e. DEMs) can be explored with GIS prior to the selection of appropriate measurements. The GIS technology was designed specifically for landscape topologies in real cartographic mapping, and it has been adopted by ecologists for their surveys. Jernvall and his colleagues (1999, 2000)

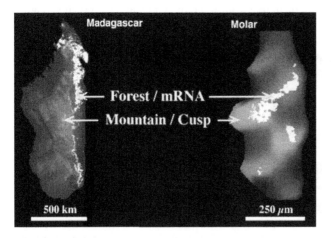

Figure 4.9 Explicit mapping analogy made by Jernvall in his GIS reconstructions of gene expression in the rodent molar (Courtesy: J. Jernvall).

have adapted this technology to measure the development of mammalian teeth. Moreover, they can place on the digital morphology the various gene expression patterns they have acquired by *in situ* hybridization. In these papers, Jernvall and colleagues have demonstrated that changes in gene expression in enamel knots (the signaling center for cusp formation) cause the morphological differences between mouse and vole molars (as seen by the DEMs). Expression of *fgf4*, *Shh*, and *p21* prefigures changes in morphology, and the spatial distribution of these genes' expression causes the subsequent location of cusps. This remarkable conclusion is predicated on extremely sophisticated gene expression and fate mapping correlations. The analogy of geographic mapping to gene expression mapping was made explicitly by Jernvall in his presentation at the SICB meeting in January 2001 (Figure 4.9).

Conclusion

Mapping is a critical part of contemporary developmental biology. Both fate maps and gene expression maps summarize new data, and organize it in a way that relates it to other data and incites new research. Gene expression mapping has been especially important in providing evidence for evolutionary developmental biology. It not only summarizes the developmental genetic evidence in the field, it also serves the functions of relating the new data to classical data (thereby linking a new science to an established science) and of showing the importance of evo-devo to evolutionary biology. New procedures in monitoring gene expression patterns, localizing dyes in certain cells, and graphically representing these patterns have made mapping of paramount importance to developmental biology.

While the "mapping analogy" may be a strained and not altogether appropriate metaphor for gene linkage studies, it is appropriate and even explicit in both the procedures and in the uses of gene expression data in developmental biology. Gene expression maps and fate maps have provided some of the best conceptual tools for the advancement of the new discipline of evolutionary developmental biology.

Acknowledgments

Funds for the completion of this work were from the Swarthmore College Faculty Research Support Grant and from National Science Foundation grant IBN-0079341. I wish to sincerely thank the organizers of this conference and the Max-Planck-Institut für Wissenschaftsgeschichte for the travel funds and hospitality. Special thanks to Ms Antje Radeck for her assistance in the preparation of this manuscript, and also the American Philosophical Society, Philadelphia, PA, for their permissions to use these correspondences.

Note

1 It is interesting that Sturtevant (1913) does not use the term "map" in his original paper describing the technique. Rather he refers to his diagram as a "linear array," in keeping with his using the mathematical relationships between the six X-linked genes to

demonstrate the correctness of Morgan's view of genes as linearly arranged on the chromosome. In the revised edition of *Mechanism of Mendelian Heredity* (Morgan *et al.* 1923), Morgan still does not use the term "maps," even though the frontispiece shows the known *Drosophila* genes arranged linearly on the four chromosomes. He refers to these figures as "diagrams" (Morgan *et al.* 1923: 61), but shows the correlation between "map-distance and crossover values" (Morgan *et al.* 1923: 170–1). In his 1926 *The Theory of the Gene*, there is an index entry for "map of the chromosome," but the text still does not use this term. Rather, the "map distance" is given in the legend to figure 19, and it is problematized by quotation marks. Indeed, his text (Morgan *et al.* 1923: 22) discusses "the 'distance' at which any two pairs of elements lie with respect to each other." Not only does Morgan problematize the word "distance" with quotation marks, but he goes on to say that this information allows one to construct "charts of the elements in each of the linkage groups." Again, he is not using the term "maps." I suspect that "maps" had too many connotations of Castle's (1919) three-dimensional model of the gene, which explicitly called Morgan's diagrams "maps" (Castle 1919: 26), and which compared Morgan's maps unfavorably to his own "models." Castle's three-dimensional model of the gene denied crossing over, and until Castle's theory was completely demolished, Morgan may not have wanted to use the map analogy.

Bibliography

Averof, M. and Patel, N.H. (1997) "Crustacean Appendage Evolution Associated with Changes in *Hox* Gene Expression," *Nature*, 388: 682–6.

Bauer, D.V., Huang, S., and Moody, S.A. (1994) "The Cleavage Stage Origin of Spemann's Organizer: Analysis of the Movements of Blastomere Clones before and during Gastrulation in *Xenopus*," *Development*, 120: 1179–89.

Beldade, P., Brakefield, P.M., and Long, A.D. (2002) "Contribution of Distal-less to Quantitative Variation in Butterfly Eyespots," *Nature*, 415: 315–18.

Brakefield, P.M., Gates, J., Keys, D., Kesbeke, F., Wigngaarden, P.J., Monteiro, A., French, V., and Carroll, S.B. (1996) "Development, Plasticity, and Evolution of Butterfly Eyespot Patterns," *Nature*, 384: 236–42.

Burke, A.C., Nelson, A.C., Morgan, B.A., and Tabin C. (1995) "Hox Genes and the Evolution of Vertebrate Axial Morphology," *Development*, 121: 333–46.

Castle, W.E. (1919) "Is the Arrangement of the Genes in the Chromosome Linear?" *Proceedings of the National Academy of Sciences USA*, 5: 25–32.

Cohn, M.J. and Tickle, C. (1999) "Developmental Basis of Limblessness and Axial Patterning in Snakes," *Nature*, 399: 474–9.

Conklin, E.G. (1905) "The Organization and Cell Lineage of the Ascidian Egg," *Journal of the Academy of Natural Sciences of Philadelphia*, 13: 1–119.

Constantini, F. (2002) Available at <http://cpmcnet.columbia.edu/dept/genetics/kidney/movies.html> (accessed January 30, 2003).

DoubleTwist (2000) "Indispensable in Uncharted Territory," *Science*, 290: 240.

Gaunt, S.J. (1994) "Conservation in the Hox Code during Morphological Evolution," *International Journal of Developmental Biology*, 38: 549–52.

Gilbert, S.F. (2000) *Developmental Biology*, 6th edn, Sunderland, MA: Sinauer Associates.

Gont, L.K., Steinbeisser, H., Blumberg, B., and De Robertis, E.M. (1993) "Tail Formation as a Continuation of Gastrulation: The Multiple Tail Populations of the *Xenopus* Tailbud Derive from the late Blastopore Lip," *Development*, 119: 991–1004.

Hamburger, V. (1988) *The Heritage of Experimental Embryology: Hans Spemann and the Organizer*, Oxford: Oxford University Press.

Hecksher-Sørensen, J. and Sharpe, J. (2002) *3D Confocal Construction of Gene Expression in Mouse*, Available at <http://genex.hgu.mrc.ac.uk/Collaborations/Sorenson_Sharpe/home.html> (accessed January 30, 2003).

Hirth, F. and Reichert, H. (1999) "Conserved Genetic Programs in Insect and Mammalian Brain Development," *BioEssays*, 21: 677–84.

Huxley, J. and De Beer, G.R. (1934) *The Elements of Experimental Embryology*, Cambridge: Cambridge University Press.

Jernvall, J. and Selänne, L. (1999) "Laser Confocal Microscopy and Geographic Information Systems in the Study of Dental Morphology," *Palaeontologia Electronica*, 2(1): 18, 905KB. Available at <http://www-odp.tamu.edu/paleo/1999_1/confocal/issue1_99.htm> (accessed January 30, 2003).

Jernvall, J., Keranen, S.V., and Thesleff, I. (2000) "Evolutionary Modification of Development in Mammalian Teeth: Quantifying Gene Expression Patterns and Topography," *Proceedings of the National Academy of Sciences USA*, 97: 14444–8.

Keller, R.E. (1976) "Vital Dye Mapping of the Gastrula and Neurula of *Xenopus laevis*: II. Prospective areas and morphogenetic movements of the deep layer," *Developmental Biology*, 51: 118–37.

Kidney Development Database, coordinated by Jonathan Bard and James Davies. Available at <http://golgi.ana.ed.ac.uk/kidhome.html> (accessed January 30, 2003).

Kishimoto, Y., Lee, K.H., Zon, L., Hammerschmidt, M., and Schulte-Merker, S. (1997) "The Molecular Nature of Zebrafish Swirl: BMP2 Function is Essential during Early Dorsoventral Patterning," *Development*, 124: 4457–66.

Lane, M.C. and Smith, W.C. (1999) "The Origins of Primitive Blood in *Xenopus*: Implications for Axial Patterning," *Development*, 126: 423–34.

Loredo, G.A., Brukman, A., Harris, M.P., Kagle, D., LeClair, E.E., Gutman, R., Denney, E., Henkelman, E., Murray, B.P., Fallon, J.F., Tuan, R.S., and Gilbert, S.F. (2001) "Development of an Evolutionarily Novel Structure: Fibroblast Growth Factor Expression in the Carapacial Ridge of Turtle Embryos," *Journal of Experimental Zoology/Molecular and Developmental Evolution*, 291: 274–81.

Manzanares, M., Wada, H., Itasaki, N., Trainor, P.A., Krumlauf, R., and Holland, P.W. (2000) "Conservation and Elaboration of Hox Gene Regulation During Evolution of the Vertebrate Head," *Nature*, 408: 854–7.

Meinhardt, H. (2002) "The Radially Symmetric Hydra and the Evolution of the Bilateral Body Plan: An Old Body Became a Young Brain," *BioEssays*, 24: 185–91.

Monmonier, M. (1991) *How to Lie with Maps*, Chicago, IL: University of Chicago Press.

Morgan, T.H. (1926 revised edition) *The Theory of the Gene*, New Haven, CT: Yale University Press.

Morgan, T.H., Sturtevant, A.H., Muller, H.J., and Bridges, C.B. (1923) *The Mechanism of Mendelian Heredity*, revised edition, New York: Henry Holt.

New England BioLabs (2001) "Mapping the Human Genome," *Science*, 291: 1176. (Other citings are possible; but this is in the special human genome issue of *Science* and was placed directly facing the main review of the human genome.)

Riddle, R.D., Johnson, R.L., Laufer, E., and Tabin, C. (1993) "Sonic Hedgehog Mediates the Polarizing Activity of the ZPA," *Cell*, 75: 1401–16.

Schierwater, B. and Kuhn, K. (1998) "Homology of Hox Genes and the Zootype Concept in Early Metazoan Evolution," *Molecular Phylogenetics and Evolution*, 9: 375–81.

Sherwood, D.R. and McClay, D.R. (2001) "LvNotch Signaling Plays a Dual Role in Regulating the Position of the Ectoderm–Endoderm Boundary in the Sea Urchin Embryo," *Development*, 128: 2221–32.

Shubin, N., Tabin, C., and Carroll S. (1997) "Fossils, Genes, and the Evolution of Animal Limbs," *Nature*, 388: 639–48.

Slack, J.M.W., Holland, P.W.H., and Graham, C.F. (1993) "The Zootype and the Phylotypic Stage," *Nature*, 361: 490–2.

Sordino, P., van der Hoeven, F., and Duboule, D. (1995) "Hox Gene Expression in Teleost Fins and the Origin of Vertebrate Digits," *Nature*, 375: 678–81.

Spemann, H. and Mangold, H. (1924) "Induction of Embryonic Primordia by Implantation of Organizers from a Different Species," in B.H. Willier and J.M. Oppenheimer (eds) *Foundations of Experimental Embryology*, New York: Hafner.

Spemann, H. and Schotté, O. (1932) "Über xenoplatische Transplantation als Mittel zur Analyse der embryonalen Induction," *Naturwissenschaften*, 20: 463–7.

Srinivas, S., Goldberg, M.R., Watanabe, T., D'Agati, V., al-Awqati, Q., and Costantini, F. (1999) "Expression of Green Fluorescent Protein in the Ureteric Bud of Transgenic Mice: A New Tool for the Analysis of Ureteric Bud Morphogenesis," *Developmental Genetics*, 24: 241–51.

Streicher, J., Donat, M.A., Strauss, B., Sporle, R., Schughart, K., and Müller, G.B. (2000) "Computer-Based Three-Dimensional Visualization of Developmental Gene Expression," *Nature Genetics*, 25: 147–52.

Sturtevant, A.H. (1913) "The Linear Arrangement of Six Sex-Linked Factors in *Drosophila*, as Shown by Their Mode of Association," *Journal of Experimental Zoology*, 14: 43–59.

Tung, T.C., Wu, S.C., and Tung, Y.Y.F. (1962) "The Presumptive Areas of the Egg of Amphioxus," *Scientia Sinica*, 11: 82–90.

US Department of Energy, "Know Ourselves," Available at <http://www.ornl.gov/hgmis/publicat/tko/04a_img.html> (accessed January 30, 2003).

Vodicka, M.A. and Gerhart, J.C. (1995) "Blastomere Derivation and Domains of Gene Expression in the Spemann Organizer of *Xenopus laevis*," *Development*, 121: 3505–18.

Vogt, W. (1929) "Gestaltungsanalyse am Amphibienkeim mit örtlicher Vitalfärbung. II. Teil Gastrulation und Mesodermbildung bei Urodelen und Anuren," *Wilhelm Roux' Archiv für Entwicklungsmechanik der Organismen*, 120: 384–706.

Winklbauer, R. and Schürfeld, M. (1999) "Vegetal Rotation: A New Gastrulation Movement Involved in the Internalization of the Mesoderm and Endoderm in *Xenopus*," *Development*, 126: 3703–13.

Wray, G.A. and Raff, R.A. (1990) "Novel Origins of Lineage Founder Cells in the Direct-Developing Sea Urchin *Heliocidaris erythrogramma*," *Developmental Biology*, 141: 41–54.

5 Mapping the worm's genome

Tools, networks, patronage

Soraya de Chadarevian

Since the 1960s, when Sydney Brenner introduced the small free-living nematode *Caenorhabditis elegans* as new model organism for the study of development, the worm has been the object of extensive mapping projects. Linkage maps of its genes were followed by cell lineage maps, by detailed maps of its nervous system and by a complete physical map of its genome. In 1998, the complete sequence of its genome was announced. It was the first multicellular organism of which the full sequence was known. In connection with the other maps, the sequence, freely available to researchers on the web, established *C. elegans* as a key organism for biological research. The worm-sequencing project further served as a pilot for the human genome project. The success story of *C. elegans*, recognized by last year's Nobel award to Sydney Brenner and two of his early collaborators on the project, John Sulston and Bob Horvitz, testifies to the key role of maps in current biological practice. This point continues to be obscured by the fact that mapping is often perceived as a service function or lower status activity, the recent euphoria surrounding the sequencing of the human genome notwithstanding (de Chadarevian 2000; Ankeny 2001). At the same time, the close succession of the worm-mapping efforts, which often involved the same people, invites comparison of the different projects. Are they all part of the same 'mapping culture' or can we discern differences? And how do the different maps relate to each other?

This chapter engages with the genetic mapping efforts, including work on the linkage map, the physical map of the genome and the genome sequence. On the *C. elegans* database, developed as part of the sequencing project, a simple click on the mouse allows users to move from a locus on the genetic linkage map to its representation on the physical map and on to the sequence of the corresponding gene or, vice versa, from the molecular to the functional representation. In this powerful representation, the different maps appear as an integrated series of pictures at increasing resolution. What this unified picture does not readily give away is the successive history of the mapping projects, the different experimental cultures that sustained them and the work necessary to make the maps appear as the logical succession of one another.

Establishing *C. elegans* as an organism for genetic analysis

Having settled on *C. elegans* as new research organism for the study of development, Brenner, working at the Medical Research Council Laboratory of Molecular Biology in Cambridge (LMB), started work on the nematode, a hermaphrodite with occasional males, by chemically inducing mutations and isolating and characterizing mutants.[1] He focused on behavioral mutants, mainly those defective in movement, with the aim of matching mutations with structural alterations in the nervous system. His idea was to genetically 'dissect' the structure of the nervous system and its development in much the same way as had been done for biosynthetic pathways in bacteria. His first paper on the worm presented the results of eight years of research. During this time, he had analyzed and genetically characterized about 300 mutants. In extensive crossing experiments, performed with the help of his assistant, Muriel Wigby, he had identified 100 genes. They were found to lie on six linkage groups, which corresponded to the six chromosomes of the worm. Next to lists and tables illustrating the findings, the paper contained a genetic map, which recorded the relative distances of the genes based on recombination experiments (Figure 5.1) (Brenner 1974).

Brenner's interest in maps dated back to the very beginning of his research career. The project he then pursued and which got him hired to Cambridge was to match the fine scale mutation map of a gene as championed by Seymour Benzer for phage, with the amino acid sequence of the corresponding protein. The hope was to get clues on the structure of the genetic code, which governed gene function. Although nothing came out of these attempts, Brenner remained heavily committed to structural approaches in genetics (de Chadarevian 1996, 2002).

The linkage map established *C. elegans* as an organism suitable for genetic analysis. New loci were added to the map as more and more researchers started working on the worm. Horvitz especially took care of the map during his postdoctoral fellowship in Brenner's laboratory. Eventually, the mechanism of updating the map was more formalized and the administration of the map moved from Cambridge to the Caenorhabditis Genetics Center at the University of Missouri, where the strain collection was also kept. By that time, many more groups working on the worm were active in America than in Europe. Periodically a new map, containing an increasing amount of annotations, was published. Thus the map was a communal, but—in contrast to the drosophila mapping effort in Thomas Hunt Morgan's laboratory fifty years earlier—not a concerted effort by the worm community.[2]

It should be noted that at the beginning Brenner was ambivalent about other people working on 'his' organism and the project was not necessarily planned to become big, but from the 1970s he started actively recruiting people. Most researchers working on the worm spent a period in Brenner's laboratory before setting up their own group. This gave the group a special coherence, an aspect often stressed by worm workers themselves. Like other organism groups, the community relied on common services. Besides, the central stock collection supplementing the map, these included a newsletter, the *Worm Breeders' Gazette*, and regular meetings.

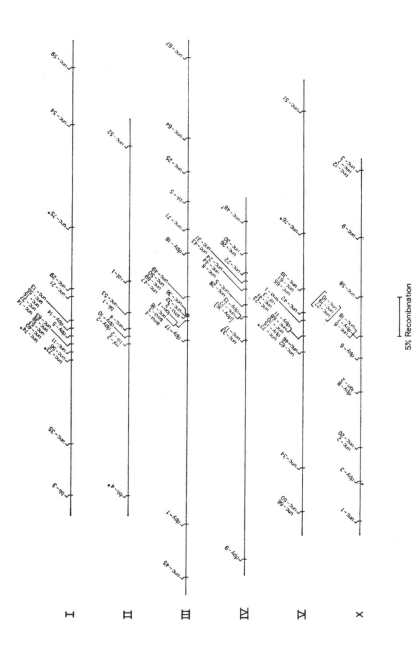

Figure 5.1 Genetic map of *C. elegans* showing six linkage groups corresponding to the six chromosomes.

Source: *Genetics* 77 (1974), p. 88, fig. 2. Copyright: The Genetics Society of America 1974.

Linkage maps record spatial relationships in only one dimension, and can therefore be seen to only represent a sequence rather than a proper map. However, scientists regularly refer to the linear representations of genes as 'maps.'[3] Since their invention by researchers working on the fly in Morgan's laboratory at Columbia in the 1910s, linkage maps have been used to characterize mutants as well as to manipulate genes to produce new mutants (Kohler 1994). But despite all speculation and increasing knowledge on the actual arrangements of genes on chromosomes, the loci on the linkage maps remained abstract entities. Distances between loci were (and still are) operationally defined in crossing experiments. They indicate relative, not actual distances on the chromosomes. With the development of recombinant DNA technologies, linkage maps nonetheless became tools to actually 'find' genes, much like geographical maps help to find places. A first gene, relating to the muscle protein myosin and thus to movement impaired mutants in *C. elegans*, was isolated and sequenced in the early 1980s. Other genes followed. 'Finding' genes became easier once a physical map of the *C. elegans* genome became available. Far from rendering the linkage map obsolete, the new map continued to rely on it. In this process the linkage map itself gained new meaning.

A new map

Work on the new map was started by a small group of researchers in Brenner's laboratory in Cambridge in the early 1980s. In 1986, a paper in the *Proceedings of the National Academy of Sciences* reported progress (Coulson *et al.* 1986). The plan was to achieve an overlapping collection of cloned DNA fragments, which covered the whole genome. The fragments were held permanently as frozen clones in a 'library' from where they could be called up. The 'fingerprint' or restriction digest data of each clone was stored in a computer database to facilitate comparison with other clones. Indeed, while the genetic map was still drawn up by hand, the development of computers and their ready availability in the laboratory were prerequisites for the physical mapping project. The main aim of the new map was to provide a tool for the worm community to facilitate the cloning of genes and to foster communication between the different laboratories engaged in this work. This also helped to make it accepted in the community.

Two years after the interim report, the adoption of a new technique, which allowed the cloning of much larger fragments in yeast, instead of bacteria, greatly accelerated the task. The technique was developed for the parallel but independent effort to map the considerably smaller yeast genome. It was imported to the worm mapping effort by Robert Waterston, a worm researcher working in the same institution of the yeast people engaged in developing the technique at the University of Washington in St Louis. Like most researchers working on the worm, Waterston had passed a training period in Brenner's laboratory. He had come back to Cambridge on a sabbatical when the physical mapping effort using bacterial clones was in full swing, but problems with the method were becoming apparent. Learning about the new cloning method developed at St Louis, he perceived at once that it would help dealing with the problems encountered at Cambridge. The *C. elegans* mapping effort developed into an international collaboration between the St Louis group headed by Waterston, and the Laboratory of Molecular Biology at Cambridge—a decision taken to avoid competition as much as anything else. Making use of new communication technologies data and information on the map in construction—much of it anyway digitalized—was communicated between Cambridge and St Louis via bitnet. To send the complex data the available transmission programs had to be stretched to the limits. With time, the number of laboratories receiving the same information grew (Figure 5.2). By 1988, the physical map of the worm's genome was virtually complete.

The physical map represented a completely new digital and physical (in form of the frozen clones) representation of the genome. Despite this fact, the production and functioning of the map built closely on previous mapping projects of the worm as well as on the structure of the worm community.

For John Sulston, the main researcher on the project, this was not the first experience with mapping. Before embarking on the genome mapping effort, he had traced the complete cell lineage of *C. elegans*. The ambition 'to identify every cell in the worm and trace lineages' was part of Brenner's original research proposal on the worm.[4] Despite this early declaration of intent, the plan to produce a full cell lineage seems to have taken form only gradually. At the beginning, several people who had joined Brenner to work on the worm were assigned to study different parts of the lineage. Division was organized according to organ systems. Possibly the plan was to eventually combine the pieces. However, not much came out of it until Sulston who, together with Horvitz, had been working on the larval cell lineage of the worm, decided to go for the full lineage. By that time he had gained considerable skill in tracing developing cells in the transparent body of the

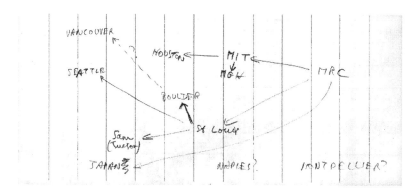

Figure 5.2 Map indicating the expanding network of laboratories receiving information regarding the construction of the physical map of *C. elegans* via bitnet [*c.*1990] (Courtesy: John Sulston).

worm under the microscope. This gave him the confidence that the work was do-able. Later he described his decision thus:

> At a certain point you say, well, if I go on doing this for another four hundred days I will have done it all, and I did think that at a point, yes. I calculated how long it was going to take and it was going to be a year and a half of look-ing down a microscope every day, twice a day, and following lineages for that time, four hours in the morning, four hours in the afternoon, that would do the whole lineage I figured.[5]

In the event the lineage took two and a half years to complete. The full cell lineage of *C. elegans* was published in 1983 (Sulston *et al.* 1983). Tracing cell line-ages was an established practice in embryology, but *C. elegans* became the first organism of which a full lineage became available. It permitted unprecedented precision in experimental manipulation. A single cell, the 'fate' of which was known exactly, could be ablated with a laser beam, and the effect of the manipu-lation studied in detail. The cell lineage also permitted the isolation and charac-terization of developmental mutants. Together with the complete map of the worm's nervous system, established in the same laboratory and published a few years later, it attracted many more people to work on the organism. By the late 1980s, around 100 laboratories around the world, although concentrated in America, Europe, and Japan, had adopted *C. elegans* as research organism; by the mid-1990s, the number had grown to 150.

Fingerprinting and matching DNA probes required completely different skills. Yet for Sulston who was trained in chemistry and whose first project on the worm had been to establish its DNA content (Sulston and Brenner 1974), embarking on the cell lineage was the more unusual step. The experience of his first mapping project nonetheless remained decisive for Sulston. It not only gave him confidence that with perseverance 'big' and apparently unattainable projects could be concluded, but also that drawing up maps and providing tools for research could be a rewarding and worthwhile activity. The special structure of the *C. elegans* community fostered this view. His account of the decision to take up the physical mapping project confirms these points:

> I heard Matt Scott talk about the work he had done mapping the drosophila genome in the region of…a homeo complex which controls structures from the anterior to the posterior of the body…and I thought that this was a huge amount of work and perhaps by doing the same amount of work but in parallel over the whole genome [of *C. elegans*], then I might be able to develop the big map of the whole thing and it really just excited me very much that this could be done because, I suppose that's an important motivation, not because it would tell anyone anything about the biological mechanism, but because the worm desperately needed this for a tool, at the time, to clone the genes.[6]

Sulston's main collaborator on the physical map, Alan Coulson, was also an experienced mapper. He began his career as assistant to Fred Sanger, head of the protein chemistry division at the LMB, when he was starting out on DNA sequencing. On Sanger's retirement, he was advised by Brenner to get in touch with Sulston who was just embarking on the project of mapping the genome. Coulson's experience in separating nucleic acid fragments on polyacrylamide gels as well as in handling large numbers of samples were pivotal to the mapping project. Coulson's collaboration with Sulston continued the longstanding interactions between Sanger and Brenner and their respective collaborators regarding mapping and sequencing techniques, first of proteins, then of nucleic acids. In 1962, the two groups had moved together in the newly established LMB in Cambridge. As Brenner then saw it, Sanger invented the techniques while the molecular biologists found the problems to apply them to (de Chadarevian 1996, 2002).

In addition to the previous mapping experiences of the researchers on the project, the success of the physical mapping effort relied heavily on the existence of a detailed linkage map of the worm, by then counting around 500 loci, and the interest of worm workers in sequencing the genes of functionally described mutants. Indeed, this was presented as main justification to begin the project. Genes for which loci and sequences were known helped both in the construction of the map, that is in linking the DNA clones, and in aligning the physical and the linkage map. As explained earlier, because of the non-linear scale on the linkage map, the alignment of the two maps was not trivial. However, only the alignment, accompanied by the attachment of gene markers, made the physical map a useful tool for researchers keen to sequence the genes of their mutants.

The construction of the physical map changed the meaning of the linkage map in at least two ways. First, the linkage map acquired a new function as mapping tool. Second, the alignment with the physical map meant that the distances on the linkage map could now be translated into physical distances. Loci on the linkage map were still defined in recombination percentages, but the same mutations could now also be materially located on specific DNA fragments. Since recombination experiments furnish only the relative distance but not the absolute order of genes, the location of genes on the physical map could serve as a check and in some instances did indeed lead to changes in the linkage map. This confirms the non-triviality of the alignment process. Representations showing the two maps running parallel reflected the new use and meaning of the lineage map (Figure 5.3).

In the early 1990s, the physical map itself was put to new use as tool for the most ambitious of all the mapping efforts: the attempt to map every single nucleotide in the worm's genome. The sequencing project relied on a new armamentarium and produced new representations of the genome. The scale of support and the visibility of the project as well as the organizational and institutional structures marked a radical shift to the earlier mapping efforts of the worm. Yet the 'philosophy' of the project remained the same. Worm researchers saw sequencing as the next logical step in their ongoing attempt to understand the

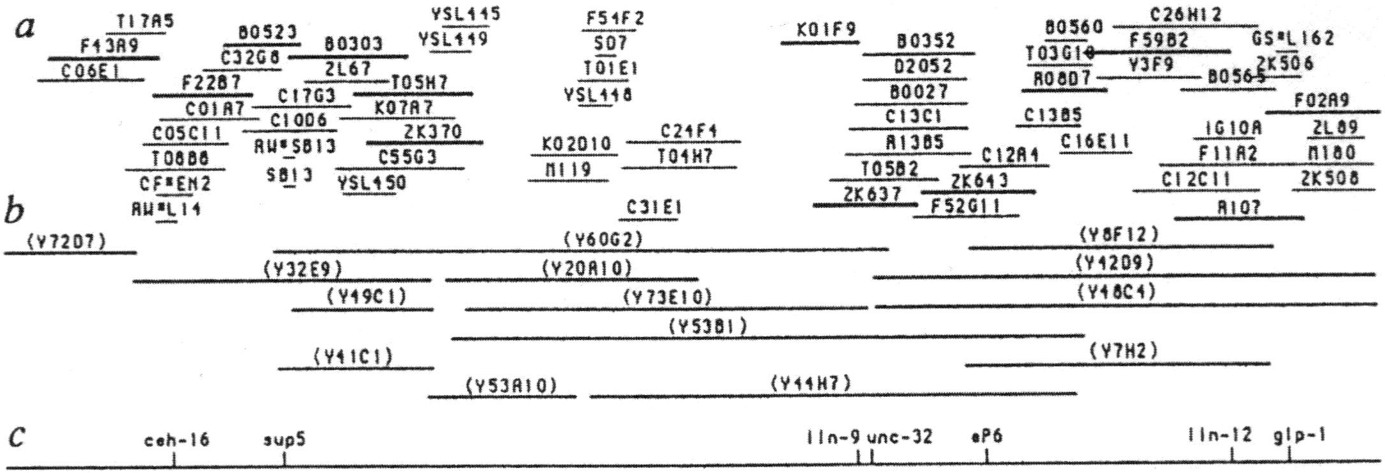

Figure 5.3 Section of physical map of *C. elegans* genome aligned with linkage map. Sequencing began in the same region: (a) set of short overlapping clones; (b) set of large overlapping clones; (c) linkage map showing genes and markers known to lie in the same region.

Source: Sulston *et al.* 1992: p. 37, fig. 1. Copyright: Nature Publishing Group 1992.

molecular machinery of the tiny worm and as providing the most powerful tool for research yet.

Raising the stakes

It is difficult to ascertain when exactly the plan came up to use the physical map of *C. elegans*, not just to find genes but as a tool to sequence the whole genome. Certainly, the same year in which Sulston and Coulson reported progress on the physical map of *C. elegans* (Coulson *et al.* 1986), Sulston also attended the conference convened by molecular biologist Robert Sinsheimer, then Chancellor of the University of California at Santa Cruz, to discuss the feasibility to sequence the human genome. By that time Sulston was still 'all *C. elegans*' and not at all interested in the human genome.[7] However, from the beginning, the project to sequence *C. elegans* became closely linked to developments concerning the human genome project.

A deal was struck in 1989. By that time it had become 'apparent that…complete sequencing of the genome might be both feasible and desirable' (*C. elegans* Sequencing Consortium 1998: 2012). Coulson and Waterston had presented the nearly complete physical map of *C. elegans* at the international worm meeting in Cold Spring Harbor (Figure 5.4). James Watson who had recently been elected as director of the newly created National Institutes of Health Human Genome Project was duly impressed. In a meeting after the workshop, the researchers signed an agreement that sequencing of the worm would be performed as a pilot of the human genome project (as researchers put it at the time: the human genome was the price to pay for the worm[8]). The American and British group

Figure 5.4 The complete map of the worm's genome displayed at Cold Spring Harbor in 1989, convincing Watson that sequencing of the genome should be attempted (Photograph by Alan Coulson; Courtesy: John Sulston and Georgina Ferry).

would collaborate. A grant of three million dollar would allow them to sequence a first stretch of three million base pairs and prove the feasibility of the project. The National Institutes of Health would fund the American part of the operation and half of the British. The Medical Research Council, which had funded the work on the map, was to pay the rest. The cooperation between the two funding agency in this project was seen by many as a 'miracle.'

Several approaches to sequencing were discussed, but in the case of *C. elegans* it was soon agreed that sequencing would make use of the available map. Clones were cut down into sub-clones to be amenable to the available sequencing techniques and a combination of 'shotgun' or random and directed sequencing techniques was adopted for each sub-clone. Proceeding clone by clone made it possible to divide up the project—for instance between St Louis and Cambridge. It also gave the flexibility to change when new methods and machines would become available. In addition, finished sequence data would become available while the sequencing project was still in progress and researchers could start working with it.

It should be noted that, unlike with the physical mapping project, the plan to sequence the whole genome met with resistance from within the worm community. Sulston recalls,

> The mapping was cheap and nobody paid any attention until it worked and then they said 'hooray, it's helping us get our clones and that's fine.' The sequencing was . . . different.[9]

Part of the dispute regarded the way the physical map should be used. Some held that the map should be used not to 'walk' the whole genome, but as a tool to target particular areas. The dispute on methods and on the usefulness of sequencing the whole genome was fueled by fears that too much money would be detracted from other worm projects. However, these fears soon lost their basis. Indeed, new funds were coming in at a rate no one could have expected.

At the end of the pilot phase the researchers from St Louis and Cambridge presented a continuous 2.2 million base pair long sequence from a gene-rich cluster of chromosome III of *C. elegans*. The result confirmed that large-scale sequencing with the available methods was feasible. In the *Nature* article reporting the results, the authors indicated that increase in scale; automation and computerization would lead to further significant improvements. This was an important projection since sequencing the whole genome on the current scale of operation would still have taken another 50 years to complete. The researchers also reported that sequencing had led to the identification of new genes and the discovery of 'intriguing features' of the genome, including putative gene duplications and a variety of other repeats with potential evolutionary implications (Wilson *et al.* 1994).

Events, however, had already overtaken the feasibility study. By the time the *Nature* article appeared, a deal was in place for the construction of a dedicated sequencing center at the outskirts of Cambridge. Money did not come from the government, but from the Wellcome Trust, the biggest medical charity in Europe. The American part of the operation saw a similar scale-up.

Developments regarding these events were reported and commented upon in the media and in the news section or editorials of *Nature* and *Science*.[10] With the human genome sequencing project gearing up, a private American investor, Frederick Bourke, actively searched for top genetic researchers to join a new company he was planning to set up in Seattle to pursue sequencing on an industrial scale (Anderson and Aldhous 1992). Sulston and Waterston, attracted by the prospect to scale up their operation—a step impossible to take in their current institutional set up—seriously considered carrying the *C. elegans* sequencing project to the company (Figure 5.5). This upset Watson, a stout defender of the international and public character of the human genome-sequencing project, which included the sequencing of model organisms. In a series of high-level conversations he convinced the Wellcome Trust to enter the fray. It did so in big style, laying the ground for a new Genome Campus, the biggest in Europe, at Hinxton near Cambridge and appointing John Sulston as its director (Aldhous 1993). In the center—the future Sanger Centre, now the Sanger Institute—sequencing would be performed on an industrial scale. Of course, the *C. elegans* sequencing project alone could not warrant such a big investment. Thus, although sequencing the worm remained on the agenda, the main effort of the new center was to be focused on the human genome. The center would also help with the yeast genome, which was sequenced by a consortium of European laboratories, joined by an American group, and start the sequencing of human pathogens. The European Bioinformatics Institute, an outstation of the European Molecular Biology Laboratory in Heidelberg, was set to move to the same site. In close collaboration with the sequencing center it was to provide fast and easy access to the sequencing information for the research community. The move to the same site of the UK Medical Research Council's Human Genome Research

Figure 5.5 Cramped conditions in Sulston's laboratory at the LMB in Cambridge where *C. elegans* genome sequencing began (Photograph by Robert Waterston; Courtesy: John Sulston and Georgina Ferry).

Centre, which coordinated research and maintained clone libraries and databases for the British genome community, was also discussed.

The scale-up in sequencing at the Sanger Centre—and similarly at the Genome Sequencing Center in St Louis—was quite dramatic. Sulston had already organized sequencing at the LMB in a semi-industrial way, aiming at high output and cost-effectiveness. However, if work at the LMB was organized around two automatic sequencing machines, their number soon increased more than hundred fold (Figure 5.6).

Other steps in the procedure were also automized. Initially group leaders were hired and each team worked on a given stretch of DNA from start to finish. This had also been the way work was organized at the LMB. Increasingly, however, work was divided by tasks with separate teams working on specialized tasks. The Sanger Centre advertised for unskilled workers 'to undertake routine laboratory work in DNA sequencing.' The qualifications required were 'manual dexterity and the ability to follow written instructions.'[11] Most of those hired were just over 16 years old. Researchers and technicians were joined by computer specialists. Parallel to the *C. elegans* sequence other sequences were churned out. As soon as available, sequences were subjected to a series of computer programs aided by human assessment to provide initial interpretation. They were then sent to the European Bioinformatics Centre on site and immediately made public. A luminous board at the entrance of the Sanger Centre displayed the sequence as it was released. The sequence information was also placed in the database built up parallel to the sequencing project and modestly named ACeD ('a *C. elegans* database'). Besides the annotated sequence, the database contained the physical

Figure 5.6 Sequencing machines at the Sanger Centre at Hinxton.
Source: The Wellcome Trust. Medical Photographic Library, London.

and genetic maps, both of which were still continuously updated, and references to strains and literature. It thus combined much available knowledge about the worm in a single site.

The essential completion of the *C. elegans* sequence, the first of a multicellular organism, was announced at a press conference in December 1998. The physical map, an equally pioneering achievement, had hardly been considered worth such public attention. The announcement was followed by a special section in *Science*. In the main article the *C. elegans* consortium, composed of over 400 researchers from Cambridge and St Louis, underlined the value of the sequence not only for *C. elegans* researchers but for biologists in general (*C. elegans* Sequencing Consortium 1998).

The *C. elegans* consortium honored its agreement when, after completion of the worm genome, or even parallel to it, it switched its attention to the human genome.[12] Following the approach established with *C. elegans*, the choice was made to proceed along a refined physical map. The motivation was much the same: to allow for multi-center cooperation. The use of the map distinguished the approach of the Sanger Centre, and more generally of the publicly funded sequencing effort, from that of Craig Venter, director of Celera Genomics, who proceeded by whole-genome shot gun method, which made no use of a map and, Venter claimed, was faster and cheaper, but did not allow portioning off the work. Besides the use of the map, the approach of the two fiercely opposed teams differed in the policy towards data release (Sulston and Ferry 2002). While the publicly funded team put the information on the web as soon as it became available—a policy Sulston had always insisted upon, even when commercial funding for *C. elegans* sequencing was discussed—Celera kept the data back with the intent of exploiting it commercially. When Venter's data became available the map became again an issue of contention between the two groups with the consortium accusing Venter that he could not have achieved his results without recurring to the physical map drawn up by the publicly funded initiative.[13]

With the sequence and the ACeDB database *C. elegans* provided a toolbox, which surpassed that of more established organisms for genetic research. Drosophilists could belittle the physical map of the worm with the argument that the giant chromosomes of their organism provided a more direct tool to localize genes on the chromosomes. As in confirmation of this view there had indeed been attempts to experimentally produce enlarged chromosomes in *C. elegans*. The sequence, however, not only permitted to localize known genes but also to identify *new* ones. Special software was being developed to help with the task. The information could then be used to update the linkage map, although the location of the new gene on the map could only be established through linkage experiments. Sulston recalls,

> ...there has always been a rivalry between the worm group [and the fly group] in LMB; it has always been this way. The fly group was on one side of the corridor and the worm group was on the other. And they used to argue who had the best organism. It was interesting, in the 1990s, it suddenly

became apparent to the fly community that they jolly well had to do genomes and had to sequence this thing because they were just loosing out. The worm was getting more and more advantages by having this stuff.[14]

If on the one hand, the availability of a sequence map fueled rivalries between the different organism groups, on the other hand the new tools made the boundaries between these groups more permeable. Of course, this relied in part on the fact that sequences were more directly comparable than physical maps. However, the new interest in genomics also denoted a shift in the values attached to sequence information, fueled by ever more powerful tools available for the analysis of this information. With the ACeDB database, freely available on the net, *C. elegans* was at the vanguard of this new culture.

Conclusion

Researchers producing the linkage map, the physical map, and the genome sequence of *C. elegans* relied on different technologies, representational devices, work organizations, institutional set ups, and patronage. The difference in culture was particularly marked in respect to the sequencing project. If the linkage map was started by a single researcher and the physical map represented the concerted effort of a small group of researchers, the *C. elegans* sequence was assembled on industrial lines, in a process involving several hundred people. The difference was not just a consequence of the scale of the work, but of historical contingencies and the new interests in sequencing, bound up with plans to sequence the human genome, of which *C. elegans* became a pilot. However, despite these differences there were also continuities between the different mapping efforts. Most importantly, the production and interpretation of each consecutive map relied on the preceding one. In this process, the maps acquired new uses, which affected their meaning and status as a whole. From a device to characterize mutants and genetically dissect behavior, the linkage map became a tool to assembly a physical map of the genome and to find the DNA sequence of functionally described genes. From an abstract representation of the genome the physical map became a tool for sequence assembly, while the sequence, in tandem with the appropriate crossing experiments, allowed researchers to add new unpredicted genes to the linkage map. Far from displacing the existing maps and their original functions, the construction of new maps strengthened their value. Their common representation in a database invited yet new ways to use the different maps and to relate compare, annotate and retrieve the stored information.

Another point emerges from this. Worm researchers have all along stressed the communicative and service function of maps. This supported the 'philosophy of collective endeavor' (Waterston and Sulston 1995: 10836), which has served as justification and has helped acceptance of the increasingly expensive mapping efforts. The immediate publication policy of the sequence was part of the same philosophy. The metaphors inspired by the use of geographical terms in genetics, like for instance 'genetic landscapes,' also invites contemplation rather than intervention. From the perspective of the worm, however, this may look differently. As other maps, too, each new map of the worm opened new ways of intervention and control: as linkage maps helped to construct new mutants and the lineage allowed researchers to selectively ablate single cells whose development was known, so the DNA sequence made it possible to selectively 'knock out' genes and to study the effects. A final map then would be one that makes it possible to reconstruct *C. elegans*, *in silico* as well as *in vivo*. If sequence maps can provide these tools, only time will show.

Acknowledgment

I thank Sydney Brenner, Alan Coulson, Jonathan Hodgkin, and John Sulston for generously making time to talk to me about their work on *C. elegans* as well as for providing source material.

Notes

1 On Brenner's extensive search for the right organism see Ankeny (2001); on the setting of Brenner's work on the worm see de Chadarevian (1998, 2002).
2 With the institution of the *C. elegans* database, including the linkage map, at Cambridge in the early 1990s, map keeping has moved back to Britain, where updating is now performed using special software (see below). On chromosome mapping in Morgan's laboratory see Kohler (1994).
3 Consistent with this usage, scientists regard the genomic sequence as the 'ultimate map'; Interview Sulston, Hinxton, February 16, 2001.
4 M. Perutz, F.H.C. Crick, J.C. Kendrew, and F. Sanger, 'The Laboratory of Molecular Biology. Proposals for Extension,' October 1963, Appendix I; reprinted in Wood (1988: xiii).
5 Interview Sulston, Hinxton, January 20, 1999.
6 Interview Sulston, Hinxton, February 16, 2001.
7 Ibid.
8 Interview Sulston, Hinxton, February 16, 2001. Watson had pleaded before for the inclusion of animal model systems as part of the human genome-sequencing project. However, *C. elegans* did not appear on the original list of organisms. Worm workers suspect Watson did this deliberately to sting them into action. For a first-hand account of the events see Sulston and Ferry (2002: 56–68).
9 Interview Sulston, Hinxton, January 20, 1999.
10 Access to archival material regarding at least some of the negotiations will not be available for a long time. As a charity, the Wellcome Trust does not even need to comply to the thirty-year rule which applies to public records in Britain; grant applications are generally considered as closed. For an insider account see Cook-Deegan (1994) and, more recently, Sulston and Ferry (2002).
11 Cambridge Evening News, March 3, 1994.
12 Sulston and Waterston had actively promoted sequencing the whole human genome for some years.
13 The yeast and *E. coli* genome were also sequenced clone by clone on the basis of a physical map. Yet data release was not immediate and complete. In contrast to the case of *C. elegans*, many of the participating laboratories in these projects, although

publicly funded, were small laboratories which performed sequencing as part of their own research projects. They aimed to explore the data for their own research purposes before making them public, a practice generally accepted in the research community.

14 Interview Sulston, Hinxton, February 16, 2001. More recently the worm group, in its turn, had been slow to include more functional data in its ACeDB database. This shortcoming is now met with the creation of *wormbase*, a new database developed at the California Institute of Technology, which will include this information. The drosophila sequence became available in 2000.

Bibliography

Aldhous, P. (1993) "Europe's Genomes Come Home to Roost," *Science*, 260: 1741.

Anderson, C. and Aldhous, P. (1992) "Genome Project Faces Commercialization Test," *Nature*, 355: 483–4.

Ankeny, R. (2001) "The Natural History of *Caenorhabditis elegans* Research," *Nature Review/Genetics*, 2: 474–6.

Brenner, S. (1974) "The Genetics of *Caenorhabditis elegans*," *Genetics*, 77: 71–94.

C. elegans Sequencing Consortium (1998) "Genome Sequence of the Nematode *C. elegans*: A Platform for Investigating Biology," *Science*, 282: 2012–18.

Cook-Deegan, R. (1994) *The Gene Wars: Science, Politics, and the Human Genome*, New York and London: Norton & Co.

Coulson, A., Sulston, J., Brenner, S., and Karn, J. (1986) "Toward a Physical Map of the Genome of the Nematode *Caenorhabditis elegans*," *Proceedings of the National Academy of Sciences*, 83: 7821–5.

de Chadarevian, S. (1996) "Sequences, Conformation, Information: Biochemists and Molecular Biologists in the 1950s," *Journal of the History of Biology*, 29: 361–86.

—— (1998) "Of Worms and Programmes: *Caenorhabditis elegans* and the Study of Development," *Studies in the History and Philosophy of the Biological and Biomedical Sciences*, 29: 81–105.

—— (2000) "Mapping Development or how Molecular Is Molecular Biology?" *Journal for the History ond Philosophy of the Life Sciences*, 22: 335–50.

—— (2002) *Designs for Life: Molecular Biology after World War II*, Cambridge: Cambridge University Press.

Kohler, R.E. (1994) *Lords of the Fly: Drosophila Genetics and the Experimental Life*, Chicago: Chicago University Press.

Sulston, J.E. and Brenner, S. (1974) "The DNA of *Caenorhabditis elegans*," *Genetics*, 77: 95–104.

Sulston, J. and Ferry, G. (2002) *The Common Thread: A Story of Science, Politics, Ethics and the Human Genome*, London: Bantam Press.

Sulston, J. *et al.* (1992) "The *C. elegans* Genome Sequencing Project: A Beginning," *Nature*, 356: 37–41.

Sulston, J.E., Schierenberg, E., White, J.G., and Thomson, J. N. (1983) "The Embryonic Cell Lineage of the Nematode *Caenorhabditis elegans*," *Developmental Biology*, 100: 64–119.

Waterston, R. and Sulston, J. (1995) "The Genome of *Caenorhabditis elegans*," *Proceedings of the National Academy of Sciences*, 92: 10836–40.

Wilson, R. *et al.* (1994) "2.2 Mb of Contiguous Nucleotide Sequence from Chromosome III of *C. elegans*," *Nature*, 368: 32–8.

Wood, W.B. and the Community of *C. elegans* Researchers (eds) (1988) *The Nematode Caenorhabditis elegans*, Cold Spring Harbor: Cold Spring Harbor Laboratory.

Part II

The moral and the political economy of human genome sequencing

6 Making maps and making social order

Governing American genome centers, 1988–93

Stephen Hilgartner

With two "draft" versions of the sequence of the human genome published in 2001, the mapping and sequencing of the human genome is often presented as among the major scientific successes of the late twentieth century.[1] In this celebratory context, the daunting obstacles that the Human Genome Project (HGP) faced, a mere decade before, have receded into the realm of history. But at the close of the 1980s, as political support for a concerted effort to analyze the human genome solidified, even the project's most enthusiastic supporters acknowledged that the available technology remained wholly inadequate for achieving the project's goals—producing genetic and physical maps of the human genome and the genomes of a number of model organisms, improving genome technology, and obtaining the complete nucleotide sequence of the human by 2005 (National Research Council 1988). The US genome program, which was slated to spend some $3 billion over a fifteen-year period under the auspices of the National Institutes of Health (NIH) and the Department of Energy (DOE), was predicated on significant technological advances. Indeed, to create a crude map of just a single human chromosome with the technology of the late 1980s was a massive undertaking, and sequencing whole mammalian genomes was deemed out of the question. Many scientists feared that the HGP would become a financial boondoggle, and critics contended, privately and sometimes publicly, that sequencing the human genome was a waste of scarce research funds. The central technical challenge was finding ways to increase the rate of data production, a problem that the project's supporters expected to solve through some combination of incremental improvements in existing techniques, automation, and perhaps major technological breakthroughs.[2] At the same time, the challenge of building organizations capable of mapping and sequencing on a large scale was a critical dimension of the problem. Like the rest of molecular biology (Knorr-Cetina 1999), genetic mapping was a world of small, independent laboratories, and these laboratories—while capable of engaging in focused hunts for individual human genes—were considered too small to tackle the problem of working on a genomic scale.

During the early years of the HGP, genome programs in England, France, the United States, and elsewhere adopted the strategy of building somewhat larger and better-funded laboratories focused on genome research (Jordan 1993).[3]

These laboratories, often formally or informally referred to as "genome centers," came to play a leading role in the HGP.[4] The NIH and DOE decided that genome centers were needed to realize economies of scale, and, no less important, to encourage the synergistic collaborations—among biologists, computer specialists, engineers, and others—needed to find ways to speed data production. Although the size, goals, structure, and politics of these new centers varied to a considerable extent across and within nations, in each case the act of creating them raised questions about how these new laboratories—with their greater size and productive capacity—would fit into the existing world of molecular genetics. Many critics of the genome project, as well as some of its supporters, worried that large genome centers might waste resources, hoard data, or exploit their size to outcompete smaller, more traditional laboratories. At the same time, the question of how to encourage productive exchange—both among genome centers and between genome centers and small laboratories—also grew salient: on the one hand, the project was expected to require an unprecedented degree of cooperation; but on the other hand, cooperation is problematic in the individualistic culture of molecular biology, as Knorr-Cetina (1999) points out. For these reasons, establishing effective genome centers not only required creating technological systems for manipulating DNA; it also required creating a social order governing these centers and their relations with the outside world. Put otherwise, building genome laboratories was a problem of co-production: the simultaneous creation of scientific knowledge and social order in a given domain (Jasanoff 2004). This chapter briefly examines how genome researchers in the United States worked to build laboratories that addressed these intertwined problems of technological and social order, focusing on the years 1988 to 1993. In this chapter, I limit my attention to the American context; very different patterns emerged in the United Kingdom (e.g. Balmer 1996; Hilgartner 2004) and in France (Rabinow 1999) during the same period. This account is based on extended field research in the genome mapping and sequencing community, including extensive interviewing and participant observation in laboratories, scientific meetings, and advisory committees. I begin by describing the challenges that genome mapping laboratories faced in the early years of the HGP, paying special attention to the social dimensions of these tensions. Finally, I discuss some of the strategies used to govern genome centers and make them accountable.

The challenge of mapping genomes

To understand the challenges of building laboratories capable of analyzing DNA on a genomic scale, one must examine the state of play in genome mapping during the years immediately preceding and following that official launch of the US genome project in 1990. In 1987, as the influential US National Research Council (1988) report *Mapping and Sequencing the Human Genome* was being completed, researchers had begun to create physical maps of small genomes, but only the most rudimentary genetic and physical maps of larger genomes were available. The genome of *E. coli* was the largest one mapped using restriction

Table 6.1 Estimated sizes of selected genomes
(base pairs)

Bacteriophage lambda	50,000
E. coli	4,700,000
Yeast	12,000,000
Drosophila	180,000,000
Human	3,000,000,000

mapping, and the human genome was three orders of magnitude bigger (see Table 6.1). Moreover, the molecular genetics techniques underlying genome mapping were labor-intensive and time-consuming. The assemblages of automated sequencing machines, laboratory robots, and informatics systems deployed as the genome project accelerated in the 1990s had not yet taken shape. Producing the complete nucleotide sequence (often described as highest resolution map) of large genomes with the technology of the late 1980s was deemed out of the question. Thus, the HGP was widely expected to begin by building low-resolution maps, which would immediately be useful to a wide range of biomedical research. Pilot sequencing projects would also be funded, but a shift to large-scale sequencing would occur only after substantial improvement in technology.

In light of the need to scale up significantly, genome researchers were (and remain today) preoccupied with finding ways to analyze more DNA, faster, cheaper, more easily; to manipulate bigger pieces of DNA (and thus simplify "jig-saw puzzle" problems); to drive down costs; to expand productive capacity; and to extract more information from the same amount of work. Michael Fortun (1998) does not exaggerate when he writes that an obsession with speed was con-stitutive of the HGP. Indeed, the goal of increasing the speed and efficiency of data production was integral to the early blueprints of the HGP, such as *Mapping and Sequencing the Human Genome*, which framed an ongoing commitment to this goal as one of the distinguishing characteristics of the project:

> The human genome project should differ from present ongoing research inasmuch as the component subprojects should have the potential to improve by 5– to 10–fold increments the scale or efficiency of mapping, sequencing, analyzing, or interpreting the information in the human genome.
>
> (National Research Council 1988: 2–3)

Genome researchers pursued this goal in a huge variety of ways, from modify-ing laboratory protocols, to introducing automation, to scaling up organization-ally so as to increase the number of workers engaged in data production. The variety of strategies cannot even be hinted at here; suffice it to say that a con-certed effort to increase what came to be known in the genomics community as "throughput" was a—and arguably *the*—central theme of efforts to enhance genome technology.

The National Research Council report and other visionary statements influential in defining the HGP also stressed that the project would require unprecedented sharing of data and materials among laboratories:

> The human genome project will differ from traditional biological research in its greater requirement for sharing materials among laboratories. For example, many laboratories will contribute DNA clones to an ordered DNA clone collection. These clones must be centrally indexed. Free access to the collected clones will be necessary for other laboratories engaged in mapping efforts and will help to prevent a needless duplication of effort. Such clones will also provide a source of DNA to be sequenced as well as many DNA probes for researchers seeking human disease genes.
>
> (National Research Council 1988: 76)

Thus, the project was expected not only to require technological advances but also social engineering.

Craft work

The ambitious vision of the HGP contrasted sharply with the capabilities of molecular genetics laboratories at the outset of the project. In the late 1980s and early 1990s, genome mapping was painstaking, manual work, generally performed by some combination of technicians, postdoctoral associates, and students. Some of the newly emerging genome centers used "factory" metaphors to describe their "assembly lines," but if these early genome laboratories were factories, they were factories based on craft work.[5] One chromosome mapping laboratory where I did participant observation in 1988 and 1989 provides an example. As was typical at the time, the lab had no robots for manipulating samples and no systems for automated data capture. Computers served primarily as word processors or tools for searching Genbank and other publicly available biomolecular databases. Laboratory workers recorded data in paper notebooks by hand, managed dozens of samples in individual tubes, pipetted liquids manually, and manipulated viscous radioactive liquids in plastic bags. To get a sense of the nature of this work, consider this description from my fieldnotes of just one step in a plasmid prep—one of many standard laboratory procedures used in molecular genetics:[6]

> The samples had been spinning in the Beckman centrifuge at 45,000 rotations per minute since Friday. Tuesday morning Maria and I...removed them from the centrifuge and very carefully, without shaking them, carried them to Maria's bench, where we were going to work on them. The reason that you have to be careful not to shake them, Maria explained, is that any agitation will cause the bands, which are separated out in the liquid to diffuse, and it will not be possible to pull out just the DNA you want. Maria carefully removed the centrifuge tubes from the rotor and then we slit each tube in the top with a razor. Then, using a syringe, we pierced the tube and withdrew the lower of the two bands. The bands, which had been stained

with ethidium bromide, were visible with a UV light, and we wore glasses to protect our eyes. And then we drew the bands out and discarded the needles. And then we discharged the fluid into a dialysis membrane, which we sealed by placing clamps on each end. And, after checking for leaks, we immersed the sealed membranes in a TE solution that I had mixed up...The dialysis membranes with the DNA samples inside will be left in the TE overnight and Maria and I will complete the work tomorrow. I think we'll then run the samples on the gel to estimate the DNA concentration...

(Fieldnotes 1988)[7]

Beyond simply being tedious, laboratory techniques were also uncertain to succeed. Laboratory materials could be extremely recalcitrant, as illustrated by the following account:

The gel system just wasn't working. They weren't sure why, but they believe some sort of contamination seems to have gotten into the system and is breaking up the DNA and causing all of the results to smear, and that it has affected all of the gel boxes. Tom showed me some nice pictures that were made prior to the failure of the system. He showed me the smeary pictures that had been made after the failure...He said that Jackie and Bill had run around trying to figure out what was happening. They had changed all of the buffers. They had changed the agarose. They had washed the boxes. They had switched the distilled water. They had looked in all the chemical boxes to see if there was...any crud in there, anything the wrong color or something. Jackie had talked about doing various analytical tests on some of the buffers and things, but these really hadn't, in Tom's view, been a good idea because they were too expensive, time consuming, and people didn't know what they were looking for so they wouldn't be very effective. In addition, he said that finally the strategy was, "Let's just try to wash this stuff out of the system. So we'll keep running gels, even though they aren't working."

(Fieldnotes 1989)

Not only were experiments regularly deemed "failures," but even when successful, their results often required painstaking interpretation.

In this craft work context, researchers conducted a variety of kinds of mapping projects, with different goals, using different technologies. To understand this world, it is useful to distinguish between two different kinds of work in human genetic mapping that prevailed at the time: gene hunting and genome mapping.[8] Gene hunting was aimed at identifying and isolating genes that caused the most common Mendelian disorders—such as Huntington's disease, cystic fibrosis, and various forms of muscular dystrophy—which represented promising "targets" for gene hunting. Finding such disease genes required creating high-resolution maps in the specific region of the genome where the gene was believed to lie. In contrast, genome mapping aimed to develop low-resolution maps of entire genomes, or, in the case of the human genome, of specific chromosomes. Put otherwise, gene hunting required building detailed local maps; genome mapping entailed creating lower-resolution maps with long-range continuity. In addition, most of the larger genome mapping laboratories aimed not only to produce map data, but also to develop mapping methods and demonstrate the viability of particular technological strategies. Gene hunting in the late 1980s was laborious, time consuming, expensive, and uncertain to succeed. The main strategy employed to find genes was a two-step process. Initially, scientists sought to use genetic linkage mapping to localize a gene to a general region of the genome. Subsequently, they would construct detailed physical maps of that region, with the goal of honing in on the particular gene, isolating it, and sequencing it. This process often took years of work and involved dozens of scientists and technicians. For example, the Huntington's disease gene, which proved particularly elusive, was localized in 1983 to human chromosome 4 (Gussela *et al.* 1983), but the gene was not found for a decade (Huntington's Disease Collaborative Research Group 1993; see also Nukaga 2002). The collaborative group that announced the discovery in 1993 listed 58 members, and this figure does not include some competitors who had also spent years searching for the gene. In the long run, genome mappers expected that HGP data would make it much easier to find genes. As one genome center director put it in 1991,

> Some of the disease genes that have already been isolated may in fact win some of those individuals Nobel prizes. On the other hand, in five years you probably won't even be able to get a Ph.D. thesis isolating a disease gene. You're going to have to do more with it than . . . I think we really will get to a point where everybody will say, "That's great," but it's not like your career is going to be solidified just because you got out another disease gene.
> (Interview 1991)

Genome mappers, thus, expected the value in the academic reward system of finding genes to decline, as map and sequence data transformed gene hunting into an increasingly "routine" procedure. However, despite uncertainty and controversy about intellectual property policy in this domain, interest in obtaining patents on disease genes—especially concerning genes implicated in common diseases, such as cancer—grew significantly during the 1990s, inspiring the founding of a wave of gene discovery companies.

Competition

Human disease gene hunting in the 1980s was not only arduous, it was also extremely competitive.[9] The number of "good targets" for gene hunts was small, owing to the fact that relatively few Mendelian disorders appear frequently in human populations, and in each case, competing groups of scientists raced to find the relevant gene. Researchers would invest years of effort in these winner-take-all games for scientific prestige, research funding, and potentially valuable patents.

Given the intense competition, strategic maneuvering over access to data was a salient feature of this research area. Molecular genetics laboratories produce and use a variety of data and resources, including inscriptions, materials, techniques, instrumentation, and many other entities. These many forms of data are woven together, forming complex assemblages—or data streams—that continually evolve as research proceeds. In scientific competition, data that are valuable and scarce or unique can convey a significant comparative advantage, and researchers use a wide range of techniques to control who gets access to what data, when, and under what terms and conditions (Hilgartner and Brandt-Rauf 1994a). Molecular geneticists deploy selected portions of their data streams in many kinds of transactions: not only do they publish data, but they also patent them, license them to commercial firms that resell them, "pre-release" them to corporate sponsors prior to publication, or hold them quietly in the lab. In addition, providing carefully targeted access to data plays a key role in the construction of collaborations, which are typically built not only on the exchange of ideas, but also on the exchange of the material means of scientific production. Thus, data become "chips" in bargaining with potential collaborators, and a variety of negotiations take the following general form: *I've got these DNA samples, you've got expertise with this technique, let's pool our resources and do the following project.*

However, collaborations in molecular biology are notoriously fragile, owing to concerns that the exchanges will grow uneven, to ambiguities and conflicts over what portions of a data stream are "covered" by a collaboration, and failures of the work to develop in expected directions (Hilgartner and Brandt-Rauf 1994b). In gene hunting, access to scarce DNA samples, unpublished map data, and other resources from the "leading edge" of a data stream can convey a decisive competitive advantage, so gene hunters control access very carefully. Disease gene hunters in the late 1980s and early 1990s were extremely cautious about providing access, given the high stakes and potential for winning valuable patents. However, simply holding all data in the laboratory was rarely an option. Because a gene hunt might continue for years, gene hunters had to publish papers along the way to demonstrate progress to colleagues and funders. Moreover, because the work was so labor intensive, gene hunters often had an incentive to collaborate with other laboratories, sharing or pooling data, if trusted colleagues could be found. The typical practices for controlling access included *non-release*, in which scientists keep the data in the laboratory while research proceeds; *delayed release*, in which a large temporal gap is maintained between the generation and the sharing of data; and *isolated release*, in which a laboratory provides access only to bounded portions of its data stream that cannot be easily extended (Hilgartner 1997). Some of the techniques for retaining a competitive edge, such as renaming publicly available DNA sample to conceal their identity, generated controversy, and some scientists developed reputations for being untrustworthy, opportunistic, and stingy. In the competitive atmosphere of human gene hunting, many "races" retained a sportsmanlike spirit, but in some cases, "bad blood" developed, distrust flourished, collaborations broke down, and competition became "personal." Indeed, in backstage settings, gene hunters sometimes accused their competitors of using unfair, "mafioso tactics," such as attempting to use personal connections to block the funding of competing laboratories.[10]

Like gene hunting, genome mapping was also highly competitive. Although the zero-sum competition of gene hunting was absent, the stakes were still high; they included large grants, prestige, and—probably most important to the scientists involved—the chance to play a major role in a project of great historical and scientific significance. At the end of the 1980s, researchers pursued a variety of mapping strategies, including genetic linkage mapping, restriction mapping, radiation hybrid mapping, and contig mapping (National Research Council 1988). To win the large grants needed to operate a genome center, genome mappers needed to convince people that their mapping methods showed promise, and at least ideally, demonstrate that the method could be scaled-up to produce ever-larger volumes of high-quality data. Genome center directors also needed to show that they could create and manage productive teams of researchers. Everyone fully expected that genome centers would inevitably be evaluated in comparison with one another, and, ultimately, continued funding would depend on productivity. However, competition among genome mappers was tempered by their collective interest in maintaining support for the HGP, countering its critics, and demonstrating its value to biology. Moreover, the sheer audacity of the genome project in the face of technological limitations encouraged exchanges of materials, techniques, software, and ideas. Finally, the NIH and DOE's efforts to coordinate the project aimed to constrain competition or channel it in productive directions.

Nevertheless, the cut-throat competition of gene hunting colored human genome mapping in significant ways; for the boundary between these two types of research was only firm on a purely conceptual level. For one thing, genome mapping and gene hunting used the same basic laboratory techniques, so the resources of a genome center could be deployed in a gene hunt. For another, despite their differing goals, genome mapping and gene hunting projects could contribute useful data to one another: thus, biomaterials (such as cloned DNA samples or cell lines) and maps of small regions developed for a gene hunt could be incorporated into a chromosome map; similarly, even an incomplete chromosome map could provide a crucial starting point for more detailed mapping of a specific region, such as a set of clones known to span the region of interest. In other words, data streams could readily flow across the gene hunting/genome mapping boundary. Moreover, the scientists overlapped. Indeed, most genome mappers actively participated in hunts for specific disease genes, often through collaborations with other laboratories.

The flexible boundary between genome mapping and gene hunting posed problems for the HGP by producing a strong disincentive to share data that might be relevant to a gene hunt. At scientific meetings, such as the international workshops that focused on mapping human chromosomes, researchers often eyed each other warily, carefully considering how much data to present and how much to hold back. Indeed, scientists in the field sometimes described these meetings as games in which the players silently negotiate about how close to keep their cards to their chests. Fieldnotes from a workshop of scientists studying a particular human chromosome in 1992 provide an illustration:

> I ran into Amy [a scientist] in the lobby of the hotel in the morning on my way over [to the meeting]. She...was holding a transparency in her hand...

I said "So is that your latest data?" pointing to the transparency. She said, "That's the rest of the map." She said, "I don't know whether I am going to present it yet. There are a few problems. There's a couple of questions, so it's preliminary data. It needs to be checked out, but I may decide to present it. It depends on what my competitors do. If the competition presents incomplete data, then I will present incomplete data. I'm going last, so I'll get to decide." ... Incidentally, when she gave her talk she *did* present that viewgraph.

(Fieldnotes 1992)

The disincentive for providing access to data posed particularly severe problems for some of the public databases (Hilgartner 1995a), such as the Genome Data Base (GDB), which sought to make the latest data from chromosome mapping projects available to the scientific community. As a result, the challenge of finding ways to ensure that competition and proprietary interests did not greatly delay the submission of data was a topic of ongoing discussion among the HGP leadership.

Small labs and centers

Size differences among laboratories added another dimension to the tensions over access to data. A large genome center could easily become a central player in a gene hunt, owing to the data—especially the maps and clones—in its possession. In contrast, a small gene hunting laboratory could not make decisive contributions to a human chromosome map, although it might be able to supply some useful data. This asymmetry generated a number of issues concerning collaborations between genome centers and gene hunting laboratories. Both sides could in principle benefit from a collaboration. For the genome center, such collaborations offered a chance to be part of a splashy scientific success (if the gene was found), demonstrating the value of its maps and work. For the small lab, collaborating with a genome center offered a means to obtain extremely valuable data, thus raising the odds of winning the race. But owing to asymmetries of resources, genome centers often found themselves in a strong position, able to pick and choose among eager collaborators. This situation generated opportunities for genome centers, but it also provoked criticism of them. As one center director put it:

[All] the centers are dealing with this "Why'd you choose this guy to collaborate with?" Somebody out there will justifiably maybe say, "Wait a minute, I thought you guys got some money to make this glorious map that's going to be accessible to everybody, and then you are essentially zeroing in on this specific collaboration which is obviously going to benefit those people as well as the individuals here that are working on that."

(Interview 1991)

Centers were not only in a position to pick and choose collaborators, but they were also able to influence the structure of the collaboration significantly. For example, in one case three competing groups—one from Europe, one from North America, and one from Australia—asked an American genome center to collaborate to hunt for the same disease gene:

> Center director: "We were essentially approached by all of them at almost the same time: 'Hey, we know you guys have this resource we need.' There was only one [group] ..., who shall remain nameless, who was a little more reluctant to collaborate with one of these other groups, but we made it very clear that those were the rules of the game essentially. So in that case we probably have all the credible people in sort of a consortium, so there's probably not going to be much conflict of interest."
>
> Interviewer: "And you were basically able to compel them in a sense to all work together."
>
> Center director: "Yes, that was sort of the condition, to say, 'Yes, you'll get this but we don't see any reason to work with two groups that are not talking to one another and are essentially competitors. That's lethal.' "
>
> (Interview 1991)

In the case of another disease gene, the same center took a different approach, because a consortium among competing groups proved impossible to arrange:

> We just distributed materials to all involved people. We said, "Fine, this will be our service function role. We know we have these clones. We don't want to get involved on this one in your petty arguments. Therefore, you guys go out and kill each other and may the best man win." Essentially you give equal access to those groups and let them deal with that.
>
> (Interview 1991)

Collaborations were not the only source of tension between small laboratories and the new genome centers. In the late 1980s and early 1990s, many scientists engaged in more traditional molecular biology research expressed doubts about the scientific benefits and financial costs of the HGP, especially large-scale sequencing. One line of attack charged that genome project was uninteresting work of uncertain value, while another common criticism worried that a "big science" approach to biology might concentrate resources in the hands of large centers at the expense of small, yet extremely productive, laboratories. A temporary tightening of the NIH budget focused attention on these issues especially sharply in 1990 (Cook-Deegan 1994: 168–78). That year Bernard D. Davis, along with nearly the entire faculty of the Harvard Medical School's Department of Microbiology and Molecular Genetics, called for a "reevaluation" of the HGP.

> Although all the goals of the HGP, except for the complete sequencing of the human genome, are clearly worthwhile, there is concern over its competition with other research for funds at a time of financial stringency, and doubt that its scientific benefits justify its rapid expansion and its organization in the

pattern of big science…The HGP may therefore need reevaluation…In any such reevaluation there would be no difficulty in justifying a centralized organization for the mapping, and probably for the research on methods. However, it is not obvious that these activities justify support for the HGP at level equivalent to over 20% of all other biomedical research…Our fundamental goal is to understand the human genome and its products, and not to sequence the genome because it is there.

(Davis and Colleagues 1990: 343)[11]

Critics of the HGP also appeared on Capitol Hill and mounted a letter-writing campaign questioning the urgency of the project. In the end, the fiscal year 1991 budget was reduced, although for reasons unrelated to the critics' efforts (Cook-Deegan 1994: 173). Genome project supporters defended themselves against these attacks, questioning the critics' financial calculations, offering their own alternative figures, and arguing that the project would ultimately save funds. The HGP leadership also attempted to disown the "big science" label, arguing that the project's relatively small budget and decentralized structure made it very different from such efforts as building a space station or a particle accelerator (e.g. Kevles and Hood 1992). The key Congressional supporters of the project remained committed to its completion. Nevertheless, for the leadership of the HGP, this wave of visible scientific opposition underlined the political importance of creating good relations between genome centers and the broader community of molecular biologists.

Making genome centers accountable

As we have seen before, during the early years of the HGP the challenges of creating genome mapping laboratories were formidable. The US program believed that genome centers, which could achieve economies of scale and orchestrate interdisciplinary collaboration, were needed to develop mapping systems capable of continually increasing the rate of data production. But creating such centers was not simply a matter of developing new technology or amassing larger groups of personnel; it was simultaneously a matter of creating a social order capable of governing genome centers and regulating their conduct. Because genome centers did not fit neatly into the bench-top culture of molecular biology, establishing them raised contentious issues: Would the data that centers produced promptly be made public, or would the centers hoard them? Would centers dominate smaller laboratories by drawing them into exploitative collaborations? Would centers waste precious funds on weak science? Building politically viable genome centers required providing compelling answers to these questions. In this context, the HGP leadership developed strategies for governing genome centers intended not only to make them productive and accountable, but to make them demonstrably so.

One dimension of this effort was establishing mechanisms to coordinate the research actively. The NIH plays a minimal role in managing most research that it funds; it typically publishes loosely defined program announcements, waits for investigator-initiated grant proposals to arrive, decides which proposals to support, and maintains a hands-off approach to successful proposals. But in the case of the HGP, the NIH and DOE concluded that closer interaction was needed between the people responsible for planning and implementing the project and the laboratories doing the research. The agencies set up an explicit planning process to steer the project, and in an important, precedent-setting move, developed a five-year plan in 1989 that committed the project to meet specific, quantified goals (NIH-DOE 1990).[12] NIH and DOE also self-consciously worked to create the collective understanding among project participants that the program was responsible for meeting these goals. And they created a number of instruments—such as requiring progress reports more often than annually—aimed at keeping the program on track.

Beyond these organizational mechanisms, a variety of material practices were designed to bring order to the HGP. One key move was the US genome program's decision to require all mapping centers to report data using a particular type of mapping "landmark"—the sequence-tagged site (STS). The STS concept was proposed in 1989 by four prominent American genome researchers (Olson *et al.* 1989; Hilgartner 1995b) and was quickly incorporated into the five-year plan for the HGP.[13] The project planners adopted STSs for several interconnected reasons. First, STSs offered a way to enhance the combinability of map data. In any kind of cartography, maps operate by representing the spatial relations among landmarks. During the late 1980s, genome researchers were busily making several kinds of maps, including linkage maps, restriction maps, radiation hybrid maps, contig maps, and *in situ* hybridization maps. These different maps could not be easily correlated and compared with one another. By making STSs into the standard landmark, the maps could be merged and compared, thus speeding the integration of data and enhancing quality control. Second, requiring all genome centers to use a standard landmark also created a uniform metric for measuring progress. Not only could the five-year plan express mapping goals in terms of this metric, but this metric provided a quantifiable measure to compare the output of laboratories. STSs, thus, offered, as one mapper put it, a means to "hold people's feet to the fire" (Fieldnotes 1991).

The genome project leadership also was attracted to STSs in order to simplify the problem of guaranteeing the wider scientific community access to map data. Before STSs, all genome mapping landmarks depended on specific clones; maps produced using those landmarks could only be used by people who had those clones in their possession. For example, without access to the underlying clones, one cannot test a DNA sample for the presence or absence of a conventional mapping landmark (Olson *et al.* 1989). As a result, making a clone-based map of the human genome publicly available would require setting up a system capable of distributing hundreds of thousands of clones (National Research Council 1988: 81–2). Not only would this raise logistical problems (clones must be specially packed on ice for shipping) but it would also raise the possibility that access to maps would be slowed or denied for strategic reasons.

However, STSs can be fully described as written texts, so their users need not worry about the circulation of materials. An STS landmark consists of a short string of sequence data that occurs only once in the genome. For example, the text:

Forward primer: TCCTGGCTGAAGAGGTTCC
Reverse primer: CATTATAGGGGCCACTGGG
PCR product size: 192 (bp)

denotes an STS on human chromosome 22. Given this information, any molecular biology laboratory can use the polymerase chain reaction (PCR) to test a DNA sample for the presence of this STS. Because such landmarks can be published in print or on the Internet, without requiring biomaterials to be shipped around, STSs offered a way to ease data distribution and make it more difficult to hoard map data (Hilgartner 1998). In multiple ways, then, this new kind of landmark served as an example of what Shapin and Schaffer (1985) call a "technology of trust" that simultaneously made maps comparable, created a quantifiable measure of productivity, eased the sharing of map data with the broader scientific community, and made genome centers more accountable.

Ultimately, publishing maps and making them available to the scientific community—with no strings attached—was particularly important to establishing a legitimate place for genome centers in the world of molecular genetics. For example, one center director described how his laboratory had won the hearts and minds of the relevant community:

> I think people were royally pissed off when we got the genome center grant. ...We knew there was a tremendous amount of resentment about that...I very concretely recognized that the only way to deal with the problem was through deed rather than word. So we got to [location of the meeting], and we just handed out the map. That settled all the problems. A number of people came over and said, "That was a real coup." That settled all the resentments right there, because whatever we had, we were sharing... [M]uch about this settled by the act of producing a lot of data, and making good on our claims, and then sharing it. It was very clear to me that that was the only way to deal with it.
>
> (Interview 1992)

Deeds proved very convincing. Rapid generation and release of data and biomaterials helped the HGP win the support of a growing number of "users" and shored up support for the project.

As the discussion here shows, strategies for governing genome centers were built into the policies of funding agencies, the material forms of laboratories, the social relations among laboratories, and even the maps themselves. Through such strategies, the HGP succeeded in constituting American genome centers as accountable entities, allowing the proponents of the project to assure the scientific community that these centers would give back more than they would take. But creating centers that simultaneously solved the prevailing problems of technological and social order in the American context by no means produced stability in the rapidly moving world of genomics. To be sure, finding solutions to these problems helped justify a continuing infusion of government funds and scientific talent. But the production of better genome maps and the expansion of genome databases in the first years of the 1990s generated unprecedented commercial interest in the field, contributing to the rise of new "private" genomics and, in turn, raising new problems of technological and social order.

Acknowledgment

The author acknowledges the support of the US National Science Foundation, Grant No. SES-0083414.

Notes

1 Two competing groups, the International Human Genome Sequencing Consortium and a group based at Celera, Inc., published the drafts simultaneously in the journals *Nature* (February 15, 2001) and *Science* (February 16, 2001), respectively.
2 See Fortun (1998) on the importance of speed to genomics. On automation, see Keating *et al.* (1999).
3 An alternative strategy, based on forming networks of small laboratories, was implemented by the European yeast sequencing effort (Hilgartner 1998).
4 It is important to recognize that term "genome center" was used in several ways during the initial years of the HGP. In its most restrictive usage, it referred to those laboratories that had been designated genome centers by the US funding agencies: that is, by successfully obtaining grants under specific NIH program announcements aimed at creating "genome centers" or by being one of the "genome centers" created with DOE funds at the national laboratories. However, a number of the larger laboratories engaged in genome research identified themselves using the term "center" or "genome center"; and some universities created "genome centers," sometimes partly in the hope of building the capacity needed to win federal genome center grants. Thus, in its more inclusive usage, the term designated any relatively large laboratory that aimed to become a "center" for genome research. This chapter employs the term in its more inclusive form.
5 On craft work in molecular biology, see Fujimura (1996); Knorr-Cetina (1999); see also Clarke and Fujimura (1992).
6 See Jordan and Lynch (1992) for a discussion of the plasmid prep in a variety of local contexts.
7 Excerpts from fieldnotes and interviews are presented verbatim, except for omissions marked with ellipses, insertions in brackets, and very light editing to correct grammar. All names have been changed to preserve anonymity.
8 For a textbook outlining various genome mapping and gene hunting techniques, see Cantor and Smith (1999).
9 It remains competitive today, but the focus has shifted from Mendelian disorders to complex diseases.
10 The quotations are from fieldnotes.
11 How Davis and his colleagues calculated the 20 percent figure is unclear; on the budget debate before Congress, see Cook-Deegan (1994).
12 The five-year plan published in 1990 ended up being the first of a series of five-year plans.
13 For a more detailed discussion of STSs, see Stemerding and Hilgartner (1998).

Bibliography

Balmer, B. (1996) "Managing Mapping in the Human Genome Project," *Social Studies of Science*, 26: 531–73.

Cantor, C.R. and Smith C.L. (1999) *Genomics: The Science and Technology behind the Human Genome Project*, New York: John Wiley & Sons.

Clarke, A.E. (1995) "Research Materials and Reproductive Science in the United States, 1910–1940," in S.L. Star (ed.) *Ecologies of Knowledge: Work and Politics in Science and Technology*, Albany, NY: State University of New York Press.

Clarke, A.E. and Fujimura, J.H. (1992) (eds) *The Right Tools for the Job: At Work in Twentieth-Century Life Sciences*, Princeton, NJ: Princeton University Press.

Cook-Deegan, R. (1994) *The Gene Wars*, New York: W.W. Norton & Company, Inc.

Davis, B.D. and Colleagues (1990) "The Human Genome and other Initiatives," *Science*, 249: 342–3.

Fortun, M. (1998) "The Human Genome Project and the Acceleration of Biotechnology," in A. Thackray (ed.) *Private Science: Biotechnology and the Rise of the Molecular Sciences*, Philadelphia, PA: University of Pennsylvania Press.

Fujimura, J.H. (1992) "Crafting Science: Standardized Packages, Boundary Objects, and 'Translation'," in A. Pickering (ed.) *Science as Practice and Culture*, Chicago, IL: University of Chicago Press.

—— (1996) *Crafting Science: A Sociohistory of the Quest for the Genetics of Cancer*, Cambridge, MA: Harvard University Press.

Gusella, J.F., Wexler, N.S., Conneally, P.M., Naylor, S.L., Anderson, M.A., Tanzi, R.E., Watkins, P.C., Ottina, K., Wallace, M.R., Sakaguchi, A.Y., Young, A.B., Shoulson, I., Bonilla, E., and Martin, J.B. (1983) "A Polymorphic DNA Marker Genetically linked to Huntington's Disease," *Nature*, 306: 224–38.

Hilgartner, S. (1995a) "Biomolecular Databases: New Communication Regimes for Biology?" *Science Communication*, 17: 240–63.

—— (1995b) "The Human Genome Project," in S. Jasanoff, G.E. Markle, J.C. Petersen, and T.J. Pinch (eds) *Handbook of Science and Technology Studies*, Thousand Oaks, CA: Sage Publications.

—— (1997) "Access to Data and Intellectual Property: Scientific Exchange in Genome Research," in National Academy of Sciences, *Intellectual Property and Research Tools in Molecular Biology: Report of a Workshop*, Washington: National Academy Press.

—— (1998) "Data Access Policy in Genome Research," in A. Thackray (ed.) *Private Science: Biotechnology and the Rise of the Molecular Sciences*, Philadelphia, PA: University of Pennsylvania Press.

—— (2004) "Mapping Systems and Moral Order: Constituting Property in Genome Laboratories," in S. Jasanoff (ed.) *States of Knowledge: The Co-Production of Science and Social Order*, New York: Routledge.

Hilgartner, S. and Brandt-Rauf, S. (1994a) "Data Access, Ownership, and Control: Toward Empirical Studies of Access Practices," *Knowledge: Creation, Diffusion, Utilization*, 15: 355–72.

—— (1994b) "Controlling Data and Resources: Access Strategies in Molecular Genetics," paper presented at a conference on "University Goals, Institutional Mechanisms, and the Industrial Transferability of Research," Stanford University, March 2004. Online. Available at <http://www.sts.cornell.edu/hilgartner_files/controlling_data.html>

Huntington's Disease Collaborative Research Group (1993) "A Novel Gene Containing a Trinucleotide Repeat that is Expanded and Unstable on Huntington's Disease Chromosomes," *Cell*, 72: 971–83.

Jasanoff, S. (2004) (ed.) *States of Knowledge: The Co-Production of Science and Social Order*, New York: Routledge.

Jordan, B. (1993) *Travelling Around the Human Genome: An in situ Investigation*, Montrouge, France: John Libby Eurotext.

Jordan, K. and Lynch, M.E. (1992) "The Sociology of a Genetic Engineering Technique Ritual and Rationality in the Performance of the 'Plasmid Prep,'" in A.E. Clarke and J.H. Fujimura (eds) *The Right Tools for the Job: At Work in Twentieth-Century Life Sciences*, Princeton, NJ: Princeton University Press.

Keating, P., Limoges, C., and Cambrosio, A. (1999) "The Automated Laboratory: The Generation and Replication of Work in Molecular Genetics," in M. Fortun and E. Mendelsohn (eds) *The Practices of Human Genetics, Sociology of the Sciences, Yearbook*, Boston: Kluwer.

Kevles, D.J. and Hood, L. (1992) "Reflections," in D.J. Kevles and L. Hood (eds) *The Code of Codes: Scientific and Social Issues in the Human Genome Project*, Cambridge, MA: Harvard University Press.

Knorr-Cetina, K. (1999) *Epistemic Cultures: How the Sciences Make Knowledge*, Cambridge, MA: Harvard University Press.

Kohler, R.E. (1994) *Lords of the Fly: Drosophila Genetics and the Experimental Life*, Chicago, IL: University of Chicago Press.

Lenoir, T. (1999) "Shaping Biomedicine as an Information Science," in M.E. Bowden, T.B. Hahn and R.V. Williams (eds) *Proceedings of the 1998 Conference on the History and Heritage of Science Information Systems*, Medford, NJ: Information Today, Inc.

National Institutes of Health and Department of Energy (1990) *Understanding Our Genetic Inheritance, The US Human Genome Project: The First Five Years, FY1991–1995* (DOE/ER-0452P), Washington, DC: Department of Health and Human Services and Department of Energy.

National Research Council (1988) *Mapping and Sequencing the Human Genome*, Washington, DC: National Academy Press.

Nukaga, Y. (2002) "Between Tradition and Innovation in New Genetics: The Continuity of Medical Pedigrees and the Development of Combination Work in the Case of Huntington's Disease," *New Genetics and Society*, 21: 39–64.

Olson, M., Hood, L., Cantor, C., and Botstein, D. (1989) "A Common Language for Physical Mapping of the Human Genome," *Science*, 245: 1434–5.

Rabinow, P. (1996) *Making PCR: A Story of Biotechnology*, Chicago, IL: University of Chicago Press.

—— (1999) *French DNA: Trouble in Purgatory*, Chicago, IL: University of Chicago Press.

Shapin, S. and Schaffer, S. (1985) *Leviathan and the Air-Pump: Hobbes, Boyle, and the Experimental Life*, Princeton, NJ: Princeton University Press.

Stemerding, D. and Hilgartner, S. (1998) "Means of Coordination in the Making of Biological Science: On the Mapping of Plants, Animals, and Genes," in C. Disco and B.J.R. van der Meulen (eds) *Getting New Technologies Together*, New York: De Gruyter.

7 Mapping the human genome at Généthon laboratory

The French Muscular Dystrophy Association and the politics of the gene

Alain Kaufmann

> It is true that we are atypical. Less and less of course, but it is sure that we used to be. The construction, outside any campus, either scientific or medical, of 4,000 m^2 of laboratories, giving 150 million francs to three scientists: this is not something usual. And taking this decision so quickly... and in a such innovative laboratory, with the intention of tackling to the whole genome and not chromosome by chromosome. That means: implementing automation technologies. Here you see, it was radically innovative. And so, starting from there, we have set a fashion.
>
> (Gérard Peirano, Généthon administrative Director)

Foreword—starting fieldwork

When I started my fieldwork at Généthon in February 1994, the only information I had access to about this laboratory came from newspapers and from Daniel Cohen's book *Les gènes de l'espoir* (Cohen 1993). So, like anybody from the general public, my view of what was going on there was quite naïve. Of course, I was aware of the fact that something special was happening. Being trained as a biologist as well as a sociologist and wanting to investigate some issues of the ongoing Human Genome Project (HGP), I had noticed the specific interest of this very original setting. So I decided to apply for a research grant from the Swiss National Research Fund.

In any research of this kind, the question of entering fieldwork is a crucial point. In this respect I had two opportunities. First, I was hosted by the Centre de Sociologie de l'Innovation (CSI) at the Ecole des Mines. Bruno Latour, who at this time had been engaged in some research on the Human Genome Project, agreed to host me for one year in a team that was going to become more and more involved in the study of emerging biomedical issues. Second, it happened that the American anthropologist Paul Rabinow came to Paris for six months to study at the Center for the Study of Human Polymorphism (CEPH) while I was there, invited by Daniel Cohen. We first met at CEPH and immediately decided to collaborate. Following this very informal meeting, in which other CEPH researchers participated, I found myself committed to a one-year stay in the Parisian suburbs.

I received a recommendation from CEPH to enter Généthon via the DNA bank. There, the physician in charge of this unit immediately identified the observer with a Latourian figure, presenting me to everyone as the disciple of the famous co-author of *Laboratory Life* (Latour and Woolgar 1979). My informer had read this book some years ago and thought there was a lot to be learned concerning his institution using this kind of approach. I felt a little bit embarrassed, "standing upon the shoulders of such a giant."

My situation was quite unusual since there was no demand to be studied coming from the French Muscular Dystrophy Association (AFM) or Généthon themselves. They were engaged in a tough competition with other laboratories and the question of confidentiality was considered to be crucial by AFM. At this time, there were only two unpublished sociological studies ordered by the association (Barral 1991; Barral and Paterson 1991). My first intention was to concentrate on genome mapping at Généthon but more and more I realized that what constituted the originality of this enterprise was the close collaboration between the world of genomics and the AFM world. I thus decided to integrate the study of the French association into my project.[1]

Généthon building: a socio-technical monster

Some authors like Galison and Thompson, as well as Smith and Agar, have shown how much an analysis of the location, architecture, and space organization of scientific buildings could enrich social studies of science (Smith and Agar 1998; Galison and Thompson 1999). Architecture and space have a major part in defining whether people interact in a more or less productive manner. They contribute to defining the identities of the users and the boundaries between them and the outsiders. Finally, they participate in the symbolic labeling of an institution. For someone accustomed to standard laboratory settings, the Généthon building was quite an unusual place. When you entered there, you did not see classical scientific staff wearing their white coats. The welcome desk and telephone switchboard was placed under the responsibility of a young man seated on a wheel-chair and affected by a muscular dystrophy. Before going further, you had to register with him and he would give you a visitor sticker-badge with your name written on it, unless you were working there and possessed an electronic badge that gave you access to the restricted area of laboratories and the DNA bank.

On the wall, facing the main door, there was a huge drawing depicting the path chosen by AFM to go from genetic disease to cure. These two poles, separated by many meters were connected by a network representing all sectors of French society that have to be mobilized to achieve this goal. A network that a sociologist of science would dream of following and describing (Figure 7.1).

In the first room on the ground floor, on your right, you could see a group of young children on electric wheel-chairs seated in front of Macintosh computers (Figure 7.2).

I was going to learn of the very important role that this "virtual space of movement" plays in the life of those who can't move by themselves any more. On

Figure 7.1 Entry room of the AFM building with its welcome desk and its poster entitled "AFM mode of operation" (on the right). This network describes the
multiple paths that the patients association intends to follow in order to
solve the problem of handicap and genetic diseases in their social, psychological, technological, medical, and scientific dimensions. In the upper part,
the genetic, physical, and cDNA mapping projects are indicated in the
trajectory towards the cure (Photography by A. Kaufmann).

Figure 7.2 Persons affected by muscular dystrophy using computers in a special
department devoted to these kinds of activities. Many other patients in their
electric wheel-chairs, either children or adults, or members of their families
regularly visited or worked in the building (Courtesy: Généthon/AFM).

Figure 7.3 The engineer in charge of the technological development unit with a technician, adjusting a home-made robot called Saturnin, designed to pool the Yeast Artificial Chromosome (YAC) bank containing the whole-human genome. This way, the genome could be sent, split in pieces, to any lab, on a limited number of paper filters (Courtesy: Généthon/AFM).

Figure 7.4 A section of the DNA bank with sample containers filled with liquid nitrogen. DNA from hundreds of families affected by genetic diseases and dozens of different pathologies are stored here (Photography by A. Kaufmann).

the same floor, you also found the DNA bank and the technological development unit in charge of conceiving home-made robots (Figures 7.3 and 7.4).

There was a restaurant and a cafeteria where researchers, technicians, patients and their families, people from AFM and visitors met everyday, exchanging their perception of what was going on. Here, researchers and technicians were able to see daily the people they were working for—no way for them to remain with an abstract or strictly molecular conception of genetic diseases. In the cafeteria, a large notice board displayed for everyone the latest news about the genes, which had been localized and identified at Généthon or with the money raised by the French Téléthon[2] and AFM.

On the first floor, there were AFM offices and meeting rooms for all the sectors of activities: social assistance, documentation, medical services, DNA collection campaign coordination, public relations, accounting, management, patient interest groups, direction, and presidency. There was also a conference room, which was used for AFM meetings and lab seminars with external speakers. A smaller meeting room was used by the Association and for the internal lab meetings.

On the second floor was the main part of the "gene factory." The genetic map project (headed by Jean Weissenbach), the physical map project (headed by Daniel Cohen), the positional cloning project (also headed by Jean Weissenbach)... and the famous "Salle des Marks." A Mark was a huge robotized electro-blotting device. Of course, this floor was the most visited of all the building. TV channels, newspapers, and magazines from all over the world came here to take some pictures of the robots of the Salle des Marks. Large sheets of paper were displayed in the main corridor in order to show to the lay and professional visitors the evolving mapping of the human chromosomes conducted by the Généthon staff.

On the third floor was the cDNA map project[3] (headed by Charles Auffray), the computing team and facilities, the library, and the administration and direction of Généthon. From a sociological point of view, one of the first striking features in all that was the close interaction between those very different social worlds. Before entering the place, there was absolutely no way for the analyst to imagine such hybrid configuration of actors and technical devices. Starting from a project to study the genome mapping procedures, I quickly understood that this place had to be studied as a whole in order to comprehend its social, scientific, technological, and organizational innovation. In a sense, from the point of view of social studies of science, it was an amazing opportunity to have access to such a "socio-technical monster" in which a huge part of the network of actors was active at a single place. Here, nothing was ever purified and the associations between the actors were constantly negotiated and re-defined. This place quickly gave you the strange feeling of standing at the center of a major transformation of the categories of science, technology, medicine, power, identity, and suffering in which contemporary genetics is playing a crucial role. According to Star and Griesemer, I was seemingly confronted by a new type of "boundary object" (Star and Griesemer 1989). This could explain the feeling of strangeness I experienced at the beginning of my stay.

In itself, the building location had a lot to tell us about the enterprise. It was situated in Evry, in the south suburbs of Paris, far away from the prestigious biomedical research centers. Nothing on the outside could provide a clue to what was going on inside excepting maybe the unavoidable 3-m tall DNA double helix standing in the grass. The building was bought in August 1989 from the French

Figure 7.5 Night view of the Généthon-AFM building. By showing the building this way, AFM wants to underline the special character of what's going on inside. It contributes to giving a more esthetic image of an otherwise ordinary construction (Courtesy: Généthon/AFM).

branch of the computer firm Olivetti. It represents a total surface of 8,600 m^2 with 4,000 m^2 devoted to laboratories (Figure 7.5). As Gérard Peirano, administrative Director of Généthon, told me:

> We couldn't have done all this in a traditional research or clinical setting because there you can't find any space left. [...] If you have to do that in environments which are already stabilized, then you must fight against the resistance of the medical and scientific establishment, the "mandarins," who have been there for a long time and want to protect their territory. Here we stand on a virgin territory. It is what allowed us to do the job [human genome mapping] at such a great speed. [...] The problem is that we are a little bit isolated. At least, this is what some of our scientists think.
>
> (Peirano, interview by the author, October 10, 1994, tape recorded)

Talking about the low cost of the surface, he underlined that "AFM politics doesn't consist of putting the money raised by the Téléthon in the buildings. All the money we can spare is put into research."[4]

Jean Weissenbach: from the Pasteurian tradition to genomics

After spending some days at the DNA bank in quite a clandestine manner, at the beginning of March 1994, I was given an appointment with Jean Weissenbach, who was at this time the scientific Director of the lab.[5] I briefly explained to him the purpose of my study and he immediately agreed to give me full access to installations, lab meetings, and all information I may need.[6] He concluded this formal part of the conversation by saying that "it is always a good thing to know the truth." (Fieldwork notes, March 2, 1994). He immediately turned to confidential matters saying that when he decided to come to Généthon he had a lot of trouble with the Institut Pasteur, where he came from, since they thought that what was going on at Généthon was "technology but not science." Now that the lab had proved to be successful they left him in peace, even though they had created a new research unit for him. The work, which had been done at Généthon until now, is "oriented research"[7] he said, but now they were moving again to "pure research" where you are forced to investigate in all directions, as is the case for AIDS. Bernard Barataud, President of the AFM, has well understood the issue in investing his money in genome maps. The genetic map was now going to be completed and integrated with the physical map of Daniel Cohen.

He continued by saying that the major issue was now positional cloning. They were certainly going to concentrate on genes responsible for neuromuscular diseases, a way to maintain a specificity and a coherence with the fundamental goals of the AFM. "This is a completely different situation with genome mapping where you can give the job to technicians: it's a machine that works for itself." He told me that "from a sociological point of view," the most interesting thing to study is the involvement of the non-scientific staff (technicians) and the courses in basic biology they provided to them. "Sometimes, somebody understands something and it's very gratifying." Reflecting about the future, he briefly mentioned gene therapy to say that this place was not the best one since there was no "critical mass" to start.[8]

What surprised me after this first encounter, and was going to be confirmed later, was the contrast between the discourse held by Weissenbach and the public one transmitted by the media or most of the AFM staff. For the Généthon Director, the mapping job was not a fundamental achievement but a necessary roundabout before going back to real science. This is of course a big difference with the discourse presented to other actors talking about a fundamental breakthrough in knowledge and cutting-edge research.[9] In fact, when I arrived there, Weissenbach was entering a kind of scientific "crisis,"[10] which could be qualified as the Généthon "post-mapping crisis."[11] In fact, at this time, industry and capital was entering the landscape of AFM strategy. More generally, this first meeting with one of the major figures of the genome mapping "à la française" exhibits very specific traits of the CEPH–Généthon–AFM collaboration. I quickly got the impression that those actors were in a constant reflexive attitude, aiming at managing the uncertainty and detecting what could be the strategy and the actors to mobilize for achievement of the next innovation, to win the next challenge. At this time, the two main achievements by the AFM were the Téléthon and the human genome maps that were on the way to completion, almost within the decided time schedule. So the question was: what to do next? Again, this constant reflexive attitude, which implies a peculiar mixing of scientific, managerial, and sociological insight, embedded in a "culture of emergency" stands in strong contrast with a standard laboratory or public research agency.

Mapping "à la Française"

Three main questions and criticisms about human genome mapping have been phrased by many researchers and commentators since its beginning: is it feasible? Is it useful? Is it real science?[12] In his very interesting paper, Balmer investigates this question in the context of the United Kingdom (Balmer 1996). He identifies two main "styles" of mapping. The first one is "mapping for mapping's sake," what constructing maps of whole regions of chromosomes means. The second one is "mapping to go from gene location to gene function." At this taxonomy, considering our case, we could add two more styles: "mapping at one place"[13] and "mapping within a network." Even if our post-hoc historical perspective and the recent developments of genomics seem to make those distinctions quite artificial, they remain interesting in our case. In fact, there is a double historical singularity in the French situation: the mapping was done in a laboratory financed by a patient association via the Téléthon . . . and it was done in France!

In a communication given at Harvard some years ago, Gaudillière analyses very well the manner in which the particular situation of the development of molecular biology in France framed the specific way in which the HGP was (not) implemented in this country (Gaudillière 1990). He shows that on one side stood the pasteurian tradition, insisting on the importance of research done by biologists in small research units, with no heavy technology and a strong theoretical approach. On the other side stood the movement initiated by Jean Dausset around HLA research, engaging a large network of laboratories, large computer facilities, technology consuming devices, and a close interaction between clinicians and biologists.

The long-lasting lack of real engagement of French public agencies like CNRS and INSERM in the HGP relies partially in their structural difficulties in releasing new financial supports in this emerging field. The fact that the mapping job was done mostly outside those institutions has of course much more to do with the marginal—even prestigious—position occupied by CEPH in the French biomedical landscape[14] and the culture of emergency of AFM who identified in what I call "politics of the gene" the best way to walk towards the cure.[15] I should mention, however, that Jean Weissenbach was appointed by the CNRS and that the agency didn't cause him any trouble when he decided to move to Généthon.

The CEPH–Généthon–AFM collaboration was based upon the encounter of strong and atypical personalities like Cohen, Barataud, and Weissenbach (Figure 7.6). The socio-technical world they built gave them the opportunity to invent an original scientific form of life: a mixture of openness, adventure, uncertainty, and generosity. Of course the very specific context in which this adventure took place played a crucial role in revealing these personalities who under other circumstances would have remained in a much more classical track. A special alchemy developed between these individuals with complementary interests and skills.

So, to our two previous questions "is it feasible? Is it useful?" Bernard Barataud from AFM and Daniel Cohen from CEPH answered "yes" and created Généthon which began to operate at the end of 1990. But for them the condition was going towards an industrial biology. To the question "is it real science?," the answer was

Figure 7.6 The main figures of the adventure standing in the "Salle des Marks." From left to right: Charles Auffray, Ilya Chumakov (Cohen's closest collaborator), Jean Weissenbach, Bernard Barataud, Jean Dausset, and Daniel Cohen (Courtesy: Généthon/AFM).

"we don't care!" As Jean Weissenbach mentioned at our first encounter, CEPH and AFM considered that it was an obligatory passage point on the path to genetic diseases.

Scaling up genetics: what does industrial biology mean?

Let us have a look at the following extract from the Généthon internal journal that summarizes quite well the way in which AFM tells its own history:

> Généthon was born because of the encounter of two personalities: Professor Jean Dausset, President of CEPH and Bernard Barataud, President of AFM. They rapidly came to the conclusion that the only way to obtain quicker results in the research on genetic diseases was to bring the men and the necessary financial and material resources to accomplish this task to the same place. Research about genetic diseases (estimated to be at least 6,000) was of course conducted in France before Généthon was created. They were very scattered and the resources each scientific team could obtain were limited. [...] More, it is about a titan and ant work.[16] As a matter of fact, the biggest part of the basic work, that is to say bench work, requires a lot of people and time. These are redundant and tedious sources of many errors of manipulations. So, despite the desire and the will of the scientists to tackle such a task, no team could

take the risk to undertake such an activity on the long term. No government and no laboratory, either private or public, could gather the amount of money needed to undertake this research. This is why, thanks to the money raised by the Téléthon, Généthon could be created and was given an access to an industrial dimension. Each time this is possible, the daily and tedious tasks that would normally be accomplished by a large number of people are performed by automated devices designed according to the needs of the Généthon laboratory.

(*Chroniques 100 gènes—Le journal du Généthon*, no 0, May 1992)

This confirms Balmer's observations that "mapping for mapping's sake" is generally considered by researchers as boring, time consuming, and not self-rewarding since as he shows, they are better served, from a credibility point of view, by "jumping on the gene" as soon as it shows the top of its nose. The question of scaling up molecular biology to achieve the goals of human genome mapping has been at the center of the controversy about its feasibility and cost. It was absolutely central in the AFM perspective since for Bernard Barataud, contemporary biology was still at a kind of prehistoric stage and he was looking for a different approach compatible with the emergency of curing genetic diseases. This is an account he wrote about his first visit to CEPH in 1988:

Laboratories, which "were doing something about genetics," we knew some of them. For a non-expert, a bench is a bench, genetics or not genetics: some instruments, glassware, some pipettes, the unavoidable Bunsen burner and the wooden tongs, all that in a charming disorder, evoking hard work and genius discovery. Here, it's a totally different world. [...] Under the conduct of professor Daniel Cohen, student of Jean Dausset and co-founder of the Center, we discover a concentration of machines of all kinds, connected to a forest of wires. On the color screen of a computer, I meet this day for the first time four letters which are going to give me a lot of trouble during some years: A for adenine, T for thymine, C for cytosine and G for guanine. DNA! [...] Amazed, I realize what all those people are doing: they are reading one by one some of the three billion units of information that the nucleus of each one of our cells contains: the human genome. The secret of life, of heredity. *The key to our diseases and to many others, thousands of others*. The passionate discourse from professor Cohen suddenly captures my attention. It's too long, he says. We are still at the prehistory of genetics. Do you realize, six years of hard work to identify the gene responsible for the Duchenne muscular dystrophy! [...] We have to proceed in a different way. [...] We are now able to reduce the time we need to read the genome by a factor of ten. We must not count on a miracle coming from the American or Japanese technology. We have to start now, in France. We must delegate what can be delegated to machines, free the researchers of those millions of operations. *This is not more complicated than building cars on an assembly line at the Renault factory. Your problem for the coming years is not muscular dystrophies. As long as we do not improve our knowledge of the human genome, research will take a long time.*

(Barataud 1992: 275–6. Emphases added)

One page later, the President of AFM summarizes the impact of this first visit to CEPH saying, "This is how the AFM genetic project was born. *Being unable to tackle the localization of the genes responsible for our forty diseases, we came to the idea to finance the mapping of the human genome*" (Barataud 1992: 279. Emphases added).

Since "little history" sometimes also contributes to "big history," it is worth reporting how the decision to start the operation was taken by the two main scientific figures of this adventure. Weissenbach started to be interested in the genetic map in 1989. He was impressed by the fact that the available maps were at this time of a very bad quality but that it was, though, an indispensable tool for disease gene hunting. In the spring, after reading Weber's publication about microsatellites (Weber and May 1989), he decided to call Cohen to tell him that they should undertake a pilot study. As Weissenbach was going to be on a sabbatical period, Cohen said: "you come whenever you want to CEPH." Here is the way in which Weissenbach reports a discussion that was going to have important consequences:

One night [October or November 1989], when I was working in the underground of CEPH, Cohen comes to see me. We often were talking about the fact that it would be a good thing to have some money to do genomics at a bigger scale. He was often talking to Barataud and said, "yes, yes, I got the impression that Barataud is going to take the decision to start something important and at the Ministry also, they seem very hot to start something. [. . .] But there, your method we should do it at a big scale." [. . .] And I told him yes, yes, my dream is to do 500–600 [markers]. So he said "No, no, this is not 500–600, it is 5,000–6,000!" So I said, Daniel, it is not possible, we don't have the money; it's too expensive. And he said "Don't worry about the financial problems, it is not important. We'll find the money. The only thing you should do is to make up your mind about everything you would need to build such a project. And it would even be a good thing if you could give me tomorrow a quick quantitative estimate." I spent my night to tell to myself, so we have to sequence, we have to clone, we have to genotype, we have to do everything. There were techniques we hadn't mastered and we were already making a quantitative estimation. [. . .] We didn't know where the difficulties were going to be, how many people we need, we didn't see exactly all the problems we were going to face.

(Weissenbach, interview by the author,
December 19, 1994, tape recorded)

We would like to underline here that, as we are going to see later, the estimation given by Cohen—assuming that Weissenbach's memory is correct—reflects quite precisely the effective number of microsatellite markers, which were going to be placed on the final version of the genetic map, that is to say 5,264. This maybe another example of what contributed, in the mapping projects of the

CEPH–Généthon–AFM collaboration, to make Cohen as well as Barataud particularly charismatic figures: what they promise, they deliver.

But what does industrial biology really mean? The comparison between the HGP and big science—astronautics and particle physics especially—to which it tended first to be assimilated, has been widely criticized. In fact, the major socio-technical innovation implemented at Généthon had to do with the specific ways in which humans, computers, robots, and molecular biology were combined. On another scale, according to Weissenbach, Généthon can't be considered as the first genome center in general but as the first one which was not "chromosome-specific." This is of course an important part of the French singularity which, as we showed before, can be explained by the particular historical configuration which brought together, at the right place and at the right moment, the President of a patient association concerned with diseases induced by genes potentially dispersed throughout the entire genome[17] and the CEPH heritage boosted by Cohen's ambition. There would probably have been no "French" maps if they had to be financed by a patient association affected by single locus disease such as cystic fibrosis.

Controversy around scaling up

The importance given by AFM's President to this industrial turn and Fordian imagery sometimes created controversies with and within groups of scientists and technicians. In the beginning of 1991, as they were experiencing a lot of trouble in scaling up the techniques, Barataud proposed to engage engineers from the French firm Bertin to assess and improve the work organization in the laboratory. Weissenbach expresses his skepticism regarding this operation:

> Concerning automation we were going to be forced to reduce it to the strict minimum since the time necessary for development was unimaginable. So, Barataud, in his great naïvety said: "Anyway, we have to give this problem to engineers and you'll see how easily they are going to solve the problem." I was really astonished by his position and I didn't understood Daniel [Cohen] either who said: "This is not uninteresting, it is worth it to try." [...] So they came. Me, they made me nervous from the beginning. I told Gabor [no 2 in the group] you take care of them. They spent a few days in the lab and they just made some proposals concerning work organization, that's all. They made an expertise that costed us 500,000 French francs. Frankly, what did we gain from those 500,000 francs? Strictly nothing!
>
> (Weissenbach, interview by the author,
> February 12, 1995, tape recorded)

This very radical position was shared by most of the technicians I interviewed. Some of them summarized this failed attempt to a question of measuring the time they spent in the restrooms (*sic*). But more seriously, there was a large consensus among the lab staff to consider that automation is not an easy task to perform in the realm of biology where you can't control the speed at which bacterial colonies or cells are growing. Gabor Gyapay, number two in the lab, in charge of automation and coming from the pharmaceutical industry, summarized the situation saying that "standardized methodology exists but as far as Bertin is concerned, it was ridiculous" (Gyapay, interview by the author, September 29, 1994, tape recorded).

Peirano acknowledges that the operation was globally a failure, but for him it was mainly a consequence of the fact that scientists were not really ready for automation (Peirano, interview by the author, October 10, 1994, tape recorded). He mentioned that the incentives for performance were the same in science and in industry but that the former has to work on much more unstable data and protocols which are constantly evolving. In fact, after this interview, a former engineer from Bertin was recruited by Généthon and integrated into Weissenbach's team. That was facilitated by the fact that he was a very clever person and that he had a PhD in molecular biology.

Those controversies stemming from the transition of molecular biology towards industrial organization resulted mostly from the "culture of emergency" embodied by Barataud. It reflects more generally one of the profound transformations that biology is undertaking within the HGP. It also reflects the divergent timing patterns in which the actors are evolving. The time of science is not the time of the patients. The time for research is not the time for cure. One of the striking elements of the CEPH–Généthon–AFM collaboration is that the coordination of their diverging agendas could be achieved in a productive manner. And this occurred even if, as Weissenbach said, in one of his clear-cut statements, that: "AFM has two naïveties: one regarding the [gene] therapy and the other consisting in believing that automation is going to solve all the problems of biology" (Weissenbach, interview by the author, February 12, 1995, tape recorded). This would tend to prove that even in science and technology, sometimes naïvety pays.

Recruitment and work organization

Besides the many troubles scientists and technicians were going to tackle, one crucial question at the beginning of the Généthon operation was: "Who is going to do the job?" Concerning the whole project and the three maps, at the top configuration, 160 persons were working there, of which 140 had a salary. Among them, only 15–20 were scientists. Most were senior technicians. From the point of view of the persons in charge, it was easier to hire technicians to perform such a tedious job than scientists. To complete the more repetitive tasks, for example to operate the Mark II blotting robots devoted to physical mapping, they took nine female unemployed persons—called lab-assistants—to whom they gave three weeks training. This team was put under the responsibility of a technician, also a woman.

Recruitment started in May 1990. Most of the people were very young and possessed a modest professional background. Almost everybody was recruited for a three-year period, directly after a first interview, based upon the "feelings" of Daniel Cohen and Jean Weissenbach and as a function of the constantly evolving needs of the project. Everybody considers that this particular situation played an important part in the success of this adventure. With such malleable collaborators, not yet accustomed to a conventional lab environment, it was easier to implement the suitable socio-technical innovations. To put it in a more positive way, the fact that almost everybody—scientists and non-scientists—started from the same point, that is, "learning by doing," and in close spatial interaction with the AFM environment, was a major element of this accomplishment. Everyone quickly got the feeling of taking part in an exciting and valuable adventure devoted to the well-being of humanity.

At its cruising speed, the production of the map implied a rigid sequence of operations like those at a pharmaceutical firm. In order to limit routines and maintain motivation, a rotation of the people from one task to another was intended. But of course each person became more and more competent in his/her own task and finally there was almost no rotation. For example, some lab-assistants were specialized in gel pouring or polymerase chain reaction (PCR). And there were thousands of gels to pour and millions of PCRs to perform. When I arrived there, in the finishing phase of the maps and after more than three years of hard and sometimes tedious work, a kind of lassitude was beginning to spread among some members of the staff, especially the less qualified ones. This was reinforced by the fact that some of them didn't know whether they were going to leave the lab or be engaged in future projects such as the positional cloning.[18]

This constant strategic plasticity of the CEPH–Généthon–AFM collaboration resulted at the same time in an important capacity to innovate and in many discussions by the staff about what was going to come next. Of course this situation was a consequence of the speed at which techniques in genomics were evolving. Concurrency, almost absent at the beginning of the project, was becoming more and more difficult to manage. What would be the next good hit to make? Of course the response was not necessarily going to be the same from the point of view of AFM and from the point of view of the employees they were financing. From the beginning they were told that their professional future was not clearly predictable. But they all accepted to embark on the boat and try the adventure, being told they were in any case going to improve their professional background. The simple fact that the Téléthon was the major source of financing of their salaries was in itself a structural and recurrent source of uncertainty. But it was also an important motivation to achieve their goals and prove the usefulness of their job.

Secrecy, criticism, and competition

In the beginning, even if the overall objective of the mapping operation was discussed in some of its aspects by Bernard Barataud with a few key actors of the research administration and scientific community, the scale at which it was going to be conducted remained a secret. It was not communicated officially even to the complete scientific advisory board of AFM. Only a small delegation of four representatives, presided by François Gros, was informed. The objective was, of course, to circumvent the resistance of the scientific establishment. Barataud also needed to get some insurance from the ministers in charge that they would not be opposed to the project. The Administrative Council of AFM presided by Barataud was quite easy to convince to invest 150 million francs for three years in this program. The genomics roundabout seemed quite obvious to the patients' representatives.[19] According to Weissenbach, the word "Généthon" emerged from a meeting held in Evry on a Saturday morning, in April 1990, in Barataud's office. Between June 11 and October 11, 1990, all the main lab facilities were installed in the new AFM building.

Another reason for secrecy and speed was of course that they were afraid that other labs, especially the Americans, becoming aware of the scale of the operation, would engage more resources, more quickly. Because of that, in 1991, at the annual Cold Spring Harbor meeting, near New York, Weissenbach and Cohen decided to be cautious not to elicit any suspicion about what was going on in Evry:

> In 1991 we adopted a very low profile. I showed the map of chromosome 20. The people didn't even realize what we were doing. They were laughing behind Daniel's [Cohen] back saying that Frenchies were really funny people. As a matter of fact, nobody took us seriously. It was excellent! [...] In the beginning of 1991 only few people had come to visit us at Généthon. As they told me afterwards, they didn't believe in it. At the occasion of the first meeting of Généthon's scientific advisory board, David Cox told me: "Your job is really good, very important. But Daniel, do you think he will succeed?" Obviously, with his [Cohen's] strategy, he couldn't succeed [in physical mapping] just using the fingerprints. He didn't want to hear about my [microsatellite] markers. He wanted to build his map all alone. I don't know why.
>
> (Weissenbach, interview by the author,
> December 19, 1994, tape recorded)

At the same meeting, in May 1992, the situation is totally different. Généthon results create a real shock. Weissenbach presents some preliminary data in a very restricted session. There are about 500 markers on the genetic map that at this moment is built manually, that is, with no really serious software. Cohen shows his preliminary map of chromosome 21 and some other elements of his physical mapping strategy. As many as ten different abstracts including research conducted at CEPH or Généthon are presented. Even Ellen Donis-Keller and Jim Watson are impressed.

But competition continues. Just after this conference, being aware of the fact that the Americans—in fact the NIH/CEPH collaboration—were going to publish a map with bi-allelic markers, Weissenbach sends the manuscript of the first version of the genetic map to *Science*. Reviewers are very impressed and ask for some modifications. Finally, the editor refuses the paper saying, "it hurts me to turn down such a beautiful work" (Weissenbach/Kaufmann, interview, December 19, 1994). The main argument for this refusal is the redundant character of the two maps. We'll see later that this consideration has nothing to do with a scientific evaluation of the respective quality of those tools. Weissenbach's paper was finally published by *Nature* (Weissenbach *et al.* 1992). "It's too bad for *Science*" he said, "it is the most cited genomics paper in the literature."

History repeats itself in 1994 when Jeffrey Murray from University of Iowa, head of the Cooperative Human Linkage Center (CHLC), contacts Weissenbach. He wants to publish his map in *Science*, together with the one of the French geneticist. This was perceived by the French as an attempt to phagocytose their work. Back to *Nature*, again.

Quality standards: one place is better than many

From the point of view of molecular biology, the innovation brought by Weissenbach and co-workers with the genetic map can be seen as an innovative combination of three emerging technologies: large-scale DNA sequencing (Church and Kieffer-Higgins 1988), large scale use of the PCR (Saiki *et al.* 1988)[20] and multiplex microsatellite genotyping (Litt and Luty 1989; Weber and May 1989). Most of the initial efforts at the start-up of the lab in the beginning of 1991 were devoted to scaling up the combination of those techniques usually done in small numbers and with a large amount of reagents. Not only the question of scale was important but also the constant push to optimize the processes—time and reagent amounts—to lower the cost. Eventually new techniques were developed if necessary to respond to the need for performing a large number of reactions per day.[21] For example, for safety reasons implied by the large scale of the genotyping job, radioisotopes usually attached to primers in order to reveal the presence of PCR products are replaced by horseradish peroxidase (Vignal *et al.* 1993). Luminescence replaces radioactivity.

Of course, this job couldn't have been done without CEPH preliminary expertise in large-scale mapping and especially without their reference family panel collected for the purpose of HLA studies.[22] Testing of the feasibility of the map was undertaken at CEPH by Weissenbach's PhD student Jamilé Hazan with the microsatellite mapping of chromosome 20 (Hazan *et al.* 1992). At this time, it was one of the poorly mapped human chromosomes because of its lack of polymorphic markers.

It is now well known that the challenge to map the whole genome in one place, contrasting with the decentralized process adopted in the other countries mostly for political reasons, is a major part in the success of the enterprise and the quality of the maps. Especially the genetic map that gave rise to almost no criticism and was immediately adopted by the biomedical community as the best tool for gene hunting. So, as soon as the first version of the Généthon map was published by *Nature* in 1992, the previous ones were practically abandoned by the researchers. As we saw earlier, it was, ironically, the case for the one produced by the NIH/CEPH collaboration, which was also published in 1992. So, in a sense, CEPH was competing with itself. Not even to mention other maps such as the one by Donis-Keller *et al.*, which in its time gave rise to a huge competition and controversy with other groups (Donis-Keller *et al.* 1987).[23] No other mapping technique could beat microsatellites. They were more informative and more uniformly dispersed on the chromosomes. Thanks to PCR, they were quite easy to handle for a small research group. Besides, Généthon was ready to send primers to any lab at the lowest costs providing they mentioned the Généthon contribution to their work. More than 150 laboratories in the world were using the map in this way (Figure 7.7).

The two following versions of the genetic map, published in 1994 and 1996, almost satisfied the projection made by Weissenbach (Gyapay *et al.* 1994; Dib *et al.* 1996). The question of the daily quality assessment and identification of new markers was a source of constant preoccupation in the lab while I was there. Some days they were loosing some of them because of their bad response to PCR or because linkage analysis showed they were clustered, some days they were gaining others. The obsessing question was: are we going to get enough markers for the final version? Are we going to be able to satisfy our public engagement? The finishing of the job was slightly deferred in time but the number of markers marginally exceeded the predictions: that is to say 5,264 instead of 5,000 microsatellites.

The production of the Généthon physical map led by Daniel Cohen and co-workers was confronted with totally different problems with the "domestication" of Yeast Artificial Chromosomes (YAC)[24] that was mainly conducted at CEPH. In fact, as it was discretely mentioned earlier in one of Weissenbach's citations, the physical map appeared to be largely dependent on the "anchors" constituted by the microsatellite markers of the genetic map. Its first versions were published

Figure 7.7 A map prepared by the genetic mapping team for the visit of President Jacques Chirac to show him (using colored needles) the geographical settings of the laboratories using Généthon's markers throughout the world (Photography by A. Kaufmann).

successively in 1992 and 1993, the same year as the publication of the first version of the genetic map (Bellané-Chantelot *et al.* 1992; Cohen *et al.* 1993).[25] Just before that, the genomics community got the paper about the physical map of chromosome 21 (Chumakov *et al.* 1992). It was chosen as a model for physical mapping with YACs because it was the smallest one and was associated with several important genetic disorders such as Down's syndrome and some forms of Alzheimer's disease. The last version of the CEPH–Généthon physical map was published in September 1995 in The Genome Directory, by *Nature* (Chumakov *et al.* 1995).

Service to the community

In the case the region in which a defective gene was not localized at all, Généthon also offered a genotyping facility to external researchers. They were invited to come for an average duration of three weeks, bringing their DNA samples. With the help of five female technicians and some Hamilton pipetting robots they were given two sets of 200 microsatellites primers chosen for adequately covering the 23 pairs of chromosomes. Markers were used to perform a genome-wide screening. The robots were originally conceived to perform ELISA immunological testing and were "diverted" by Généthon (Figure 7.8).

In this case it was to screen the DNA samples from the families but they also served to pool the PCR primers used to build the genetic map. Using this protocol, a researcher usually working in a standard environment could sometimes gain more than one year of hard work to localize a gene. This service was launched just before the publication of the first version of the genetic map. As soon at it was known, a lot of groups applied to come to Evry. Until the moment I was there

Figure 7.8 Hamilton robot used by the visiting teams to perform a whole-genome screening by PCR. One among many other types of robots implemented in the lab, mostly for pipetting automation (Photography by A. Kaufmann).

Figure 7.9 The notice board placed in the cafeteria displaying to employees and visitors recent articles and AFM press releases relating localizations and identifications of genes achieved with the financial support of AFM and Généthon's markers (Photography by A. Kaufmann).

between 40 and 50 different groups had come, either for a whole-genome screening or for refining a localization.

It was very important for the lab to show that in addition to its proper research upon specific genetic diseases, it was offering similar services to the whole scientific community. So anyone could benefit from the power of its genetic mapping tools. It was also very important to maintain a monitoring board and communicate online the number of genes that could be localized or identified using the maps. For this purpose a regular "tally"[26] was published as well as a lot of press releases. This way, it was possible for AFM to show to everybody that the gifts made by the public via the Téléthon served as a universal benefit. At the moment I left Evry, in March 1995, the last hunting board indicated that 124 localizations and 15 identifications had been achieved partially or totally thanks to the resources created by AFM (Figure 7.9).

In fact, both scientists and the general public, via the Téléthon, manifested their support for this policy. It gave rise to a symmetric adhesion, which transformed simultaneously the dynamics of research groups involved in localizing genes and the citizens' awareness of genomics expressing their generosity by giving money to AFM. This can be interpreted as a renewal of the pasteurian tradition of science as a public good.[27]

Computing controversies

No human genome mapping is possible without computers. A consensus emerges from my observations concerning the question of computing facilities. Reflecting a more general situation in the emerging field of genomics, almost every actor I met thought that as far as data-processing was concerned, the human, technical, and financial resources engaged in the operation were not sufficient. At the very beginning of the process, geneticists had to face the fact that, in contrast to their colleagues in the field of physics, they didn't possess any consequent equipment in terms of either software or hardware. When the mapping operation started at Généthon, they used standard linkage software to make the calculations and commercial software for databases. Weissenbach was assembling his map manually. All the chromosome markers were arranged in ordinary Excel tables. With the advancement of the project, Généthon had to implement or develop many hardware and software elements in order to cope with the enormous amount of data.

Traditional molecular biology was thinking in terms of personal computers and often home-made software. Genome mapping needed big computers, UNIX, the Internet networks, standardization and professional informaticians. In this dimension too, Généthon was a kind of pioneer. Guy Vaysseix, after discussing with Mark Lathrop at this time at CEPH and conceptor of the well-known *Linkage* software was hired as the director of the computing team. He immediately tried to introduce the rules of the big informatics into the landscape. He had many difficulties in obtaining quantitative estimations of the necessary memory space needed to do the mapping. But he told me that in this period, "they had nothing to learn, even from the Americans" (Vaysseix, interview by the author, January 20, 1995, tape recorded). For him, there was nothing special about genomics. Informatics had a universal methodology to tackle any problem: algorithmics and numerical analysis. Two conceptions of biology were beginning to confront each other: the analogical and the numerical one. And the path to the second was genomics.

In fact there was always a controversy between biologists and informaticians about the kind of informatics and informaticians that were needed. Informaticians wanted to maintain their autonomy and stay at a distance from the constant demands of the users. Molecular biologists wanted to have someone at their service to cope with the unstable environment of experimental science. In fact, they often consider informaticians as a kind of prosthesis. It was a conflict between the ancient "wet" lab and the emerging "dry" lab. As I could observe, the controversies were even reflected in the kind of mapping software that the CEPH–Généthon–AFM was developing.

In our case, these diverging conceptions were geographically materialized by the fact that the core set of the computing team—which used to number as much as 20 collaborators—had its offices not at Généthon but in the center of Paris at Place de Rungis. This situation had to do with the not very prestigious status of the Evry area, the lack of space at CEPH and more decisively the will of Vaysseix to stay at a reasonable distance from biologists who where going to call you each time they had some trouble using their computers.[28] More seriously, Vaysseix thought that software development and thinking for the long term was not possible by being immersed in an environment, which didn't understand what informatics really was. Of course, some informaticians were working in Evry but most of them came from biology, being part of the more and more numerous category of bioinformaticians.

The importance of bioinformatics was sometimes underscored by the actors. As a very reflexive bioinformatician told me, it is maybe because, for the users, as in the case of an electricity network crash, the computer contribution is really appreciated only when troubles occur. The situation has radically changed since 1990, at the beginning of Généthon. As Vaysseix says, with the coming of proteomics, computer facilities in biology could well reach or even surpass the level of what physics used to know for many years. Like many other genome centers, Généthon participated massively in building a common culture between biologists and informaticians in the hybrid field of bioinformatics.

Communicating an industrial image: the case for Mark II

The "industrial face" of Généthon played a major role not only at the operational level but also at the public relations dimension. For the AFM, it was very important for it to show to the public that they were revolutionizing biology. The Salle des Marks for example, which hosted as many as 20 Mark II blotting robots, was the central figure of this marketing operation of an industrial image. Long after they ceased to play any role in the mapping project, the machines were still put into operation for the visitors.

These machines had been constructed as prototypes for the physical mapping project with financing from AFM and a European contribution. The firm Bertin was associated with their development. They were used as blotting robots to produce YAC RFLP fingerprints. As we saw earlier, the robots at their cruising speed were operated by nine unqualified women under the supervision of a technician. Concerning the physical map, the motto was that "thanks to those machines, the job that would normally have taken twenty years was done in three." In fact they also served to produce fingerprints as a service for external groups aiming at localizing a gene. But as soon as the genetic map was out, nobody wanted to use them any more. When I came to Généthon they were only good for mapping vegetal genomes.

For Daniel Cohen and Bernard Barataud, the Salle des Marks used to be an exemplary figure. It was the exact contrary of what made Barataud crazy about biology: the wood tongs and the Bunsen burner. Almost each interview by the media was done in this huge square room painted in black and white. It was an obligatory passage point for any visitor. The space was clearly designed to give you the impression of standing in a spaceship or a genetic assembly line. Especially at night when a special lightning system gave you a feeling of being projected into a strange future. A large square, closed room painted in black, with large windows was, at the center, containing the massive power units for the Mark II robots and later the automated sequencers (Figures 7.10 and 7.11). All around this room were very long white benches. The first name that came to me when I first came in was the movie by Stanley Kubrick, *2001 The Space Odyssey*.

Many people I talked to acknowledge the fact that the Mark II, as with some other home-made robots, probably served much more for the marketing and publicity

Figure 7.10 The Salle des Mark (Mark II robots room) showing the power units (behind the windows) and 6 of the 20 original Mark II robots, which were used by Cohen's physical mapping team to produce Yeast Artificial Chromosome RFLP fingerprints (Courtesy: Généthon/AFM).

Figure 7.11 Five of the nine female technicians in charge of the Mark II robots at work. Laboratory or genetic assembly line? (Courtesy: Généthon/AFM).

than for scientific output. Although the investment to conceive and operate them was very high, it was probably worth it to do the job since for the public image, the robots and their room became the icon of what was so special about Généthon.

Communicating the maps

The production of the maps constituted a major part of the communication content made by the public relations service of AFM in the scientific domain. When I met its chief, he told me that the opportunity he was given to communicate the maps was the opportunity of his lifetime. He particularly insisted on the fact that the work done by Weissenbach was a work of art. He complained that the director of the genetic mapping team was less eager to talk to the media than his colleague Daniel Cohen. After the publicity campaign for the announcement of the first version of the physical map in 1992, achieved via the electronic and printed media, the AFM public relations service was awarded a prize. This had a lot to do with Cohen's talents and public engagement. A trait of character that made him a lot of enemies. Of course, as for the industrial image, the Téléthon was one of the privileged occasions to communicate to the public the advancement

Figure 7.12 Simplified map of chromosome 1 prepared for the visitors and displayed on the walls of the genetic mapping floor. It shows the progression of the mapping enterprise by displaying the number and the location of the microsatellite markers produced between March 1992 and February 1995 (Courtesy: Généthon/AFM).

of human genome mapping (Figure 7.12).[29] As the maps were going to be completed, the scientific content of the show increasingly shifted to gene therapy assays and vector production: the next step planned by AFM a long time before the termination of the maps.

More generally, AFM was immersed in a constant communication war in order to maintain the alignment of the network of actors they had to mobilize to achieve their goals. Each communication failure could result in a drop of the resources raised by the Téléthon. One must never forget that almost all the resources of the Association are based on the success of this event. The period preceding the show was particularly delicate since sometimes, more or less explicit attacks were performed at this moment by some persons or institutions who didn't see the growing power of the patients in a positive manner. Communicating the advancement of the mapping projects via the Téléthon—and of course the number of localized and identified genes—was very important in order to maintain public support. For the patients and their families I talked to, the maps also constitute a major achievement in the building of their new "biosocial identity."[30]

Conclusion

As I said in my introduction, it's a socio-technical monster. As is the style of this chapter I'm afraid, which is a mixture of reflexive ethnography, historical report, and sociology of science. This format was intended to recreate something of the strange atmosphere of that place at that particular moment. Of course, my descriptions and reflections don't seize the thickness of this ever moving landscape. It would have taken many more pages to achieve a complete overview resulting from a one-year investigation. Even though my discourse is mainly focused on the Généthon maps production, I hope that the reader has been made receptive to the fact that in our case, everything holds together: science, technology, medicine, public relations, citizenship, politics, and . . . the market.

In fact, in my chapter the market is quite absent. When I was there, AFM was beginning to undertake some studies and ask others to evaluate the way in which they could collaborate with private firms to start a genomic park in Evry. The idea was that without them there was no way to reach the financial and technological critical mass necessary to cure genetic diseases. This new orientation, forecasted some time ago, was of course very delicate. As Rabeharisoa and Callon have shown, one of the main innovations performed by AFM was the invention of an "intermediary discourse" (Rabeharisoa and Callon 1999) placed between the state and the market. As we saw earlier, this made the achievement of the mapping project possible and was necessary for the patient association to maintain its independence over the long term. In their case independence means the possibility to innovate and adapt to a very unstable situation. Of course, the state also had to be mobilized in order to ensure that all the "seed" that had been sowed would continue to bear its fruits. To achieve their goals, AFM invests only where nobody else can do the job.

I was fortunate to observe the end of an adventure characterized by a time of exploration and innovation, and lived the termination of an epic period. Now, on the same territory, the time has come for normalization and technocracy. History is running its course, as usual.

Today, in Evry, a *Genopole* has been created. It is the result of an association between the new Généthon III (mainly financed by AFM), the new National Sequencing Center (called Genoscope) and National Center for Genotyping (directed by Jean Weissenbach and Mark Lathrop, respectively), two big private laboratory units (Genset and Aventis, formerly Rhône-Poulenc Rorer), several public labs from four national research agencies (CEA, INSERM, CNRS, and INRA)[31] and a pool of biotech start-ups. New curricula are being offered at the Evry-Val d'Essone University centered around genomics, genetics, robotics, bioinformatics, and nanotechnologies.

The development of the actual Evry Genopole manifests one of the major achievements of the *politics of the gene* conducted by AFM. It is a very consequent contribution of the concerned people to the development of what remains one of the most impressive and surprising outputs of the HGP. Généthon might be considered, according to Star and Griesemer, as one exemplary boundary object (Star and Griesemer 1989). In this respect, the maps played a crucial role since with their specific scientific and semiotic character, they are very well-suited objects to hold all these heterogeneous actors together. The spatial nature of those tools and the geographical metaphors attached to them made possible a large number of convergent *translations* (Callon 1986) by the different actors among which the Téléthon was a central part.

In fact, some of the persons concerned by genetic diseases decided to participate massively in the production of their biosocial identities. AFM, with Généthon, widely contributed to the production of what Rabinow, paraphrasing Snow, used to identify as a "third culture" (Rabinow 1994) founded on genetics and DNA technologies. In this case, we have witnessed an original articulation between the particular needs of the families represented by AFM and the "universal" goals of the genomics community. As we have shown, this was made possible by adopting a radical "politics of the gene" embedded in a "culture of emergency" which gave rise to what remains a major historical example of "knowledge co-production" in twentieth-century science (Callon 1999).

Acknowledgments

I would like to particularly thank Paul Rabinow for the close collaboration during his stay in Paris and for the friendship it helped develop. I would also thank Christian Rebollo, at this time administrative Director of the CEPH. He has been a precious informer to both of us and also became a close friend. Thanks too to Bruno Latour and Michel Callon who arranged for me to stay at CSI during this work. Michel, together with Vololona Rabeharisoa later entered the field of biomedicine and patient associations as did many other French colleagues. Participating in such common fieldwork makes it sometimes quite difficult to identify which idea comes from whom. I'm very grateful to Jean Weissenbach to have accepted me in his group during this period, with no restriction of access to any information. His modesty and cleverness played an essential part in the success of this study. Thanks also to Jean-François Prud'homme, who was my first guide at Généthon, for his "critical enthusiasm" and his constant interest in discussing any issue, often while driving me back to Paris in his car. Thanks to Jean-Paul Gaudillière who made me benefit from his profound knowledge about the French molecular biology community and spent time discussing many issues of common interest. Thanks to the direction and presidency of AFM who took the risk of accepting a foreign observer. Thanks to all the people at Généthon, AFM, and CEPH who spent some of their precious time talking to me. Thanks to Susan Cure for correcting my English and to Paul Beaud for his constant support and the opportunity he gave me to do the research.

Notes

1   In this study, data gathering was based upon participating observation. I used fieldwork notes, tape recorded interviews, and participated in lab meetings. I also collected some internal documentation like plannings, reports, internal journals, press releases, and reviews, etc. I conducted 74 in-depth interviews with 68 persons: geneticists, molecular biologists, computer scientists, physicians, technicians, engineers, chief administrators, administrative staff, public relation officers, psychologists, and patients interest groups representatives.

2   The French Téléthon takes place each year on the first weekend of December since 1987. It is broadcast for more than thirty hours by the French TV channel *France 2* and relayed by some foreign ones. The last edition managed to collect more than 500 millions French francs.

3   I'm not going to develop here the question of this specific project which is generally not considered as a major breakthrough of the Généthon initiative. This has partly to do with the fact that it had to face the savage competition with other laboratories engaged in cDNA mapping and sequencing.

4   According to this point, I had noticed that almost every piece of furniture in the offices, even those of the directors, came from a well-known Swedish firm (*sic*). It was a large contrast to the expensive architecture and furniture one often finds in prestigious research centers.

5   He is presently director of the Genoscope—Centre National Séquençage, in Evry. When I arrived there, Daniel Cohen had already moved back to CEPH for a long time. In fact I never saw him at Généthon during my stay. First, Jean Weissenbach wanted to give me Cohen's office which was unoccupied except by Cohen's piano. But finally I was attributed a work place in the central office of the genetic mapping group where all the information about the markers was retrieved and where the daily progression of the map was scrutinized using monitoring boards. It was also there that information about markers was dispatched all over the world to collaborating teams. Quite a good place to be!

6   Concerning AFM the question of accessing the people and information remained quite ambiguous for a long time. When Jean-François Prud'homme, responsible for the DNA bank, introduced me to Bernard Barataud and tried to convince him of the interest of this study, the President told me that he couldn't make such a decision on the spot. He asked me to write a research project, which I did immediately. In the following weeks, I tried many times to get a formal answer to this document because I felt the fact of being integrated at Généthon but not formally accepted by AFM quite uncomfortable. Finally, I got the response orally by one of Barataud's close collaborators and it was positive.

7   As many other collaborators, he also very often used the term "production" to qualify the work done in Evry.

8 It's worth mentioning that Généthon is now essentially devoted to research on gene therapies and called "Généthon—Centre de recherche et d'application sur les thérapies géniques."

9 Of course, today this question looks quite obvious since recently, contrasting with the triumphal discourses of the emerging HGP, most of the observers are saying that the real job is only beginning. The prestige of genomics is already decaying in the process of being framed as the prehistory of proteomics or functional genomics.

10 In a later interview, he told me that he was in a period in which, from a scientific point of view, he was looking for a new strategy. I was impressed by the modesty and the lucidity of a man who is seen as one of the major figures of contemporary genetics.

11 Another interesting crisis, of a different kind, is exposed in Rabinow's *French DNA* (1999), which also shows the importance of the management of uncertainty in this phase of the history of biology where the categories of "science," "technology," "market" and "life" are being redefined.

12 Of course, answers to those questions are constantly evolving, as are the technologies, the identity, and the strategies of the actors engaged in the controversy.

13 Which of course reflects the strategy chosen by the CEPH–Généthon–AFM collaboration.

14 On this specific question, a good synthesis is to be found in Rabinow (1999).

15 For an overview of the process that conducted AFM to invest in scientific research and genetics before the period we describe, see Paterson and Barral (1994).

16 I found no satisfactory English translation for those two French expressions.

17 According to experts, one could count more than forty different types of muscular dystrophies.

18 It is clear that in this transition from production to real science, the profile of the staff could change a lot. More qualified persons could be necessary for a less repetitive and maybe more classical work.

19 It is important to underline that the AFM Administrative Council is composed only of patients and patients' relatives in order to be sure that the goals of the association are fulfilled.

20 For the most complete account of PCR invention and its importance for contemporary genetics, see Rabinow (1996).

21 In the starting period of the project, they probably held the world record for the number of PCR reactions they were performing.

22 A total of 8 families were selected from a panel of 40 to produce the map. For a technical description of the CEPH reference family panel and collaborative mapping organization before the creation of Généthon, see Dausset *et al.* (1990).

23 For a detailed description of the competition in human genome mapping, see Cook-Deegan (1994).

24 This work was made possible by improving the technique invented by Olson's group in St Louis. See Burke *et al.* (1987).

25 Some criticism emerged in the genomics community because of the problem known as "chimerism." It means that sometimes, a single YAC could be composed of fragments from different regions of a human chromosome. When this was the case, you couldn't be sure anymore that the contiguous arrangement of YACs you extract from your bank for searching a candidate gene and sequencing was strictly reflecting the DNA sequence on the real chromosome.

26 The French expression used by the actors is "tableau de chasse."

27 I'm grateful to Paul Rabinow for making me aware of this point.

28 As a standard user, I cannot resist mentioning that in my own university too, the main computing facilties are the only one to be situated outside the campus.

29 For a brilliant investigation of the Téléthon conception and its reception by the public, see Cardon *et al.* (1999).

30 I borrow this concept from Rabinow (1992).

31 Commissariat à l'Energie Atomique, Institut National de la Santé et de la Recherche Médicale, Centre National de la Recherche Scientifique, and Institut National de Recherche Agronomique.

Bibliography

Balmer, B. (1996) "Managing Mapping in the Human Genome Project," *Social Studies of Science*, 26: 531–73.

Barataud, B. (1992) *Au nom de nos enfants: Le combat du Téléthon*, Paris: Edition Numéro 1.

Barral, C. (1991) "Naissance et développement du mouvement de lutte contre les maladies neuro-musculaires en France," unpublished report, Convention AFM-CTNERHI, Vanves.

Barral, C. and Paterson, F. (1991) "Impact du téléthon sur les associations du secteur sanitaire et social," unpublished report, Convention AFM-CTNERHI, Vanves.

Bellané-Chantelot, C. *et al.* (1992) "Mapping the Whole Human Genome by Fingerprinting Yeast Artificial Chromosomes," *Cell*, 70: 1059–69.

Burke, D.T., Carle, G.F., and Olson, M.V. (1987) "Cloning of Large Segments of Exogenous DNA into Yeast by Means of Artificial Chromosome Vector," *Science*, 236: 806–12.

Callon, M. (1986) "Some Elements of a Sociology of Translation: The Domestication of the Scallops and the Fishermen of St Brieux Bay," in J. Law (ed.) *Power, Action and Belief: A New Sociology of Knowledge?* Sociological Review Monograph, London: Routledge and Keegan Paul.

—— (1999) "Les différentes formes de démocratie technique," *Les Cahiers de la sécurité intérieure*, 38: 35–52.

Cardon, D., Heurtin, J.-P., Martin, O., Pharabod, A.-S., and Rozier, S. (1999) "Les formats de la générosité: Trois explorations du Téléthon," *Réseaux*, 95, 17: 15–105.

Chumakov, I. *et al.* (1992) "Continuum of Overlapping Clones Spanning the Entire Human Chromosome 21q," *Nature*, 359: 380–7.

—— (1995) "The Genome Directory, Supplement to *Nature*," *Nature*, 377: 175–297.

Church, G.M. and Kieffer-Higgins, S. (1988) "Multiplex DNA Sequencing," *Science*, 240: 185–240.

Cohen, D. (1993) *Les gènes de l'espoir: à la découverte du génome humain*, Paris: Robert Laffont.

Cohen, D., Chumakov, I., and Weissenbach, J. (1993) "A First Generation Physical Map of the Human Genome," *Nature*, 366: 698–701.

Cook-Deegan, R. (1994) *The Gene Wars: Science, Politics and the Human Genome*, New York: W.W. Norton & Company.

Dausset, J. *et al.* (1990) "Centre d'Etude du Polymorphisme Humain (CEPH): Collaborative Genetic Mapping of the Human Genome," *Genomics*, 6: 575–7.

Dib, C. *et al.* (1996) "A Comprehensive Genetic Map of the Human Genome Based on 5,264 Microsatellites," *Nature*, 380, Special Issue.

Donis-Keller, E. *et al.* (1987) "A Genetic Linkage Map of the Human Genome," *Cell*, 51: 319–37.

Galison, P. and Thompson, E. (eds) (1999) *The Architecture of Science*, Cambridge: The MIT Press.

Gaudillière, J.-P. (1990) "French Strategies in Molecular Biology," paper presented at Harvard University, June.

Gyapay, G. *et al.* (1994) "The 1993–1994 Généthon Human Genetic Linkage Map," *Nature Genetics*, Special Issue, 7: 246–339.

Hazan, J. *et al.* (1992) "A Genetic Linkage Map of Human Chromosome 20 Composed Entirely of Microsatellite Markers," *Genomics*, 12: 183–9.

Latour, B. and Woolgar, S. (1979) *Laboratory Life: The Social Construction of Scientific Facts*, Berverly Hills, CA: Sage.

Litt, M. and Luty, J.A. (1989) "A Hypervariable Microsatellite Revealed by In Vitro Amplification of a Dinucleotide Repeat within the Cardiac Muscle Actin Gene," *American Journal of Human Genetics*, 44: 397–401.

NIH/CEPH Collaborative Mapping Group (1992) "A Comprehensive Genetic Linkage Map of the Human Genome," *Science*, 258: 67–86.

Paterson, F. and Barral, C. (1994) "L'Association Française contre les Myopathies: trajectoire d'une association d'usagers et construction associative d'une maladie," *Sciences Sociales et Santé*, 2, XII: 79–111.

Rabinow, P. (1992) "Artificiality and Enlightment: From Sociobiology to Biosociality," in J. Crary and S. Kwinter (eds) *Incorporations*, vol. 6, New York: Zone.

—— (1994) "The Third Culture," *History of the Human Sciences*, 7: 53–64.

—— (1996) *Making PCR: A Story of Biotechnology*, Chicago, IL: The University of Chicago Press.

—— (1999) *French DNA: Trouble in Purgatory*, Chicago, IL: The University of Chicago Press.

Rabeharisoa, V. and Callon, M. (1999) *Le pouvoir des malades: l'Association Française contre les Myopathies et la recherche*, Paris: Presses de l'Ecole des Mines.

Saiki, R.K. *et al.* (1988) "Primer-Directed Enzymatic Amplification of DNA with a Thermostable DNA Polymerase," *Science*, 239: 487–91.

Smith, C. and Agar, J. (1998) *Making Space for Science: Territorial Themes in the Shaping of Knowledge*, London: Macmillan Press and Center for the History of Science, Technology and Medicine, University of Manchester.

Star, S.L. and Griesemer, J.R. (1989) "Institutional Ecology, 'Translations' and Boundary Objects: Amateur and Professional in Berkeley's Museum of Vertebrate Zoology, 1907–1939," *Social Studies of Science*, 19: 387–420.

Vignal, A. *et al.* (1993) "Nonradioactive Multiplex Procedure for Genotyping of Microsatellite Markers," *Methods in Molecular Genetics*, 1: 211–21.

Weber, J.L. and May, P.E. (1989) "Abundant Class of Human DNA Polymorphisms Which Can Be Typed Using the Polymerase Chain Reaction," *American Journal of Human Genetics*, 44: 388–96.

Weissenbach, J. *et al.* (1992) "A Second-Generation Linkage Map of the Human Genome," *Nature*, 359: 794–801.

8 Sequencing human genomes

Adam Bostanci

The history of the Human Genome Project is a history of mapping projects. In its course, geneticists and molecular biologists surveyed the human chromosomes with cytogenetic, genetic, and physical markers. The maps featuring these land-marks were subsequently often collated or mapped onto one another, and eventually biologists began to "sequence" human DNA, a process customarily explained as mapping the hereditary material at the highest possible resolution. But even this final phase of the Human Genome Project occurred not once, but twice. In February 2001, independent research groups described preliminary drafts of the human DNA sequence in separate publications: a consortium of lab-oratories commonly known as the Human Genome Project published its draft in *Nature* (International Human Genome Sequencing Consortium 2001). The other version of the human genetic code was the product of Celera Genomics, a com-pany in Rockville, Maryland. This draft was described in the same week's issue of *Science* (Venter *et al.* 2001).

Visionary molecular biologists have always conceived of the human genome as a single natural object. Dubbing their fantastic plan of determining the sequence of its chemical building blocks as the "holy grail of genetics," they predicted that the human genome sequence would eventually become "the central organizing princi-ple for human genetics in the next century" (Waterston and Sulston 1998: 53). Arguably, the publication of two human DNA sequences threatens to spoil these aspirations. Making a virtue out of necessity, one team of scientists concluded in February 2001: "We are in the enviable position of having two distinct drafts of the human genome sequence. Although gaps, errors, redundancy and incomplete annotation mean that individually each falls short of the ideal, many of these problems can be assessed by comparison" (Aach *et al.* 2001: 856).

Relating the discoveries of the Human Genome Project and Celera Genomics as imperfect yet comparable representations of the same natural object is a com-mon move, but by no means the only way of making sense of these discoveries. Alternatively, one might conceive of the two versions of the human genome as separate objects. This is not to say that the drafts are irreconcilable in principle, but for the time being there are good reasons to speak of them as distinct objects in sci-entific practice. First, independent research groups working with different methods produced them: The Human Genome Project pieced together its version of the human genetic code in a "map-based sequencing" operation. Celera Genomics, in contrast, employed a "whole-genome shotgun." Second, the two drafts differ in ways that go beyond superficial sequence similarities and discrepancies, such as their topology. Third, it remains to be seen whether the two drafts of the human genome will be reconciled at all.

In the controversies surrounding efforts to sequence the human genome, the scientific strategies of the Human Genome Project and Celera Genomics were usually compared in terms of feasibility and cost. Each camp tried to persuade the public that its own approach was better suited for sequencing the human genome. My aim is not to address these disputes directly, but to extract from them information about the diverging scientific practices in the two projects. As one observer pointed out, from "an unbiased view, the two plans seem to be simply two methods to obtain genomic information" (Goodman 1998: 567). Hence I compare the research centers at which map-based sequencing and the whole-genome shotgun emerged and examine the social and epistemic roles genome maps came to play in these dissimilar sequencing regimes. This analysis shows that the disputes between the Human Genome Project and Celera Genomics emerged in the first instance from differences in scientific practice, which were later connected with gene patenting and the publication of human genome sequence data.

Mapping genomes

In scientific practice, mapping projects go hand in hand with the alignment of people, practices and instruments (Turnbull 2000: ch. 3). The production of coherent maps requires the introduction not only of standardized measures but also of standardized measuring practices, which, in turn, often impinge on moral and productive economies more generally. The Human Genome Project is a case in point: many biologists initially opposed the production of "global" maps of the human genome because they feared such an endeavor would undermine their scientific autonomy (Cook-Deegan 1994; Balmer 1996). Even as the program was slowly incorporated into the cottage industry of molecular biology of the early 1990s, alternative mapping and sequencing regimes were evaluated not only as experimental techniques but also as quasi-political modes of organization. As I will illustrate, genome maps themselves later came to play an important organizational role in the sequencing regime of the Human Genome Project.

Around 1990 several teams of scientists around the world were constructing "physical maps" of the human genome. This type of genome map depicts the order and distance between genomic sites at which "restriction enzymes" sever DNA. The enzyme *Hind*III, for instance, cuts DNA molecules in two wherever it encounters the sequence of chemical building blocks abbreviated by the letters AAGCTT. When DNA is exposed to *Hind*III and the resulting fragments are analyzed, AAGCTT is said to become a "landmark." Physical mapping projects thus establish the order and spacing of such landmarks within a genome of interest. The resultant "maps" comprise both an ordered collection of DNA fragments and a catalogue detailing the cleavage sites on them (Figure 8.1).

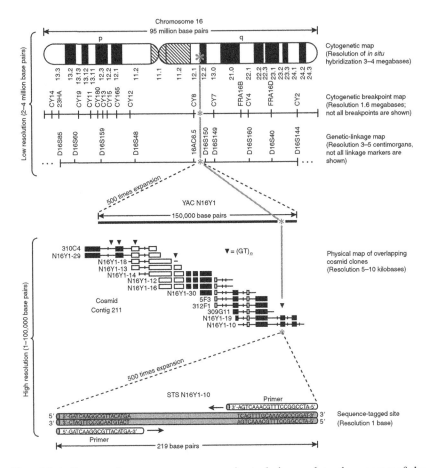

Figure 8.1 A diagram of common genome-mapping techniques relates the progress of the Human Genome Project from low to high resolution.

Source: Cooper 1994: 205, reproduced with permission by University Science Books.

Yet, the sequence AAGCTT may occur hundreds of times in a large genome; knowledge of this landmark alone is not sufficient to describe a genomic location unambiguously. Consequently, molecular biologists initially only mapped particular chromosomes or chromosomal regions and developed additional landmarks, which often varied from laboratory to laboratory. As one scientist explained, one reason for dividing the early Human Genome Project by chromosome was to preserve the structure of scientific research: "No one wanted a large monolithic organisation dictating how the information would be gathered and disseminated" (R.K. Moyzis, quoted in Cooper 1994: 111).

Others perceived this lack of standardization as divisive. Maynard Olson, a geneticist then at Washington University in St Louis, recalled that he regarded large-scale physical mapping "as a kind of tower of Babel" because it gave rise to maps expressed in "completely incompatible languages" (M.V. Olson, quoted in Cooper 1994: 123). The vision of the leaders of the Human Genome Project was to build one coherent map of the genome, which led Olson to the conclusion that, eventually all the maps "would have to be done again by whatever method proved most generic" (M.V. Olson, quoted in Cooper 1994: 123). In response to this threat, he proposed to map the genome with sequence-tagged sites (STS), a new set of landmarks that eliminated the space for local variation.

Olson defined STS as sequence targets that are "operationally unique in the human genome," that is segments of DNA that can be detected in "presence of all other genomic information" (Olson *et al.* 1989: 245). For example, the site bounded by the pair of sequence tags AGTTCGGGAGTAAAATCTTG and GCTCTATAGGAGGCCCTGAG is a unique genomic site on chromosome 17 (example from Hilgartner 1995). Notably, STS are unique in the genome because they are defined by longer sequence tags than the enzymatic cleavage sites used to construct physical maps. Another advantage of STS was that these landmarks could be described fully as information in an electronic database and, unlike physical maps, required no exchange of delicate biological materials.

Olson presented STS as a "common language" that would "solve the problem of merging data from many sources" (Olson *et al.* 1989: 245). But as Hilgartner has argued, the new landmarks also embodied a particular management philosophy for the Human Genome Project. While the chromosomes initially emerged as the natural units of its organization, STS later became the technological mode of management. STS were endorsed by scientific leaders because they represented those features of the human genome that permitted to tighten a network of dispersed laboratories by making their research commensurable and cumulative (Hilgartner 1995).

By 1996, the first global STS-map of the human genome had been assembled (Schuler *et al.* 1996). This consensus or unified map of the genome, as it also came to be known, contained more than 15,000 STS that had been mapped by an international consortium of laboratories. This map was "global" in two senses: on the one hand, it made the genome accessible as one coherent domain, rather than a collection of "local" physical maps. On the other hand, research centers anywhere in the world could henceforth work with landmarks that were conveniently available in an electronic database. STS made it possible to align existing physical maps with the consensus map. Later the Human Genome Project adopted the consensus map as an organizational framework for genome sequencing.

Sequencing

The term "sequencing" refers to a routine form of analysis that establishes the succession of the four biochemical building blocks A, T, C, and G in DNA molecules. In plain language, sequencing is commonly explained as mapping DNA at the highest possible resolution. Indeed, biologists often invoke the theme of increasing resolution in popular accounts of the Human Genome Project. The analogy makes it possible to relate the collection of genome maps that were produced in the course of the Human Genome Project as a series of ever more accurate representations, and to rationalize its history as a natural progression from mapping at low resolution to mapping at high resolution. In addition, the analogy permits to differentiate sequencing from previous mapping projects in the Human Genome Project: sequencing reveals the territory of the genome itself, not another partial map thereof. As one report predicted, the human DNA sequence would be "the ultimate map" containing "every piece of genetic information" (The Sanger Centre and The Washington University Genome Sequencing Center 1998: 1197.) But alluring as this tale of progress may be, the practical relationship between genome mapping and DNA sequencing in the Human Genome Project has been more haphazard and more interesting.

Some of the first biochemical reagents and a strategy for large-scale DNA sequencing were developed during the 1970s by Cambridge biochemist Fred Sanger, who later shared one of his two Nobel Prizes for the invention of this method. During the 1980s, his reagents were modified in such a manner that the results could be sequentially read by a computer, and commercial DNA sequencing machines became available in 1987. Since then, DNA sequencing machines have been the workhorses of all large-scale genome-sequencing projects. But no matter by how much DNA sequencing technology has improved, one fundamental limitation of this form of biochemical analysis has remained: only several hundred letters of the genetic code can be deciphered per sequencing experiment. The chemical composition of longer DNA molecules has to be re-constructed by aligning the partial sequence overlaps of DNA fragments. According to one of the first reports, this method, "in which the final sequence is built up as a composite of overlapping sub-fragment sequences, has been aptly termed 'shotgun' DNA sequencing" (Anderson 1981: 3015). Notably, both the Human Genome Project and Celera Genomics eventually used versions of the shotgun strategy to sequence human DNA. The two genome-sequencing strategies are schematically represented in Figures 8.2 and 8.3 (for a review see Green 2001).

In theory, there is no limit to the length of DNA molecules that can be sequenced by means of the shotgun method, the fundamental sequencing strategy. As long as a sufficiently large number of overlapping fragments are generated, the DNA sequence can be re-constructed by aligning overlaps. In practice, however, sequencing large genomes in this manner creates formidable challenges: data from millions of sequencing experiments has to be stored, manipulated, and assembled with special computer algorithms. Assembly is especially complicated if the genome contains repetitive sequences, which give rise to ambiguous overlaps. Ultimately, assembly may fail or require data from additional experiments to resolve ambiguities. Despite these shortcomings of the sequencing technology available in the early 1990s, some biologists established large-scale genome-sequencing projects at specialized research facilities. After all, one stated aim of the Human Genome Project was to sequence the entire human genome by 2005.

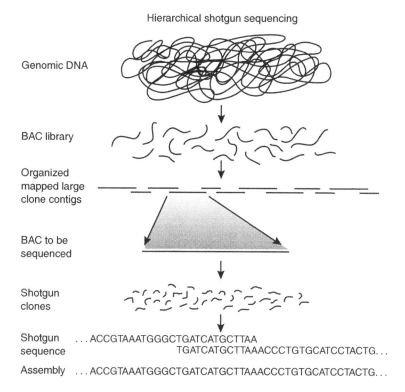

Figure 8.2 Genome sequencing: the strategy employed by the Human Genome Project
simplified the problem of shotgun assembly by sequencing mapped fragments
one at a time rather than the whole genome at once.

Source: International Human Genome Sequencing Consortium 2001, reproduced with kind permission by *Nature* and Eric Lander.

The Sanger Centre

The Sanger Centre was opened in the small village Hinxton near Cambridge
in 1993 (Fletcher and Porter 1997). Its founding director John Sulston had several years earlier constructed a complete physical map of the genome of the
small nematode worm *C. elegans* and, more recently, embarked on a project to
sequence its genome at the Laboratory of Molecular Biology in Cambridge. From
the outset, John Sulston organized this sequencing project as a transatlantic collaboration with Bob Waterston at Washington University in St Louis. Generous
funding from a biomedical charity, the Wellcome Trust, enabled Sulston to
re-locate into some disused laboratory buildings on a country estate in Hinxton, and
to expand the sequencing project. Waterston concurrently established a genome-sequencing center in St Louis. Contemporary observers hailed their cooperation as

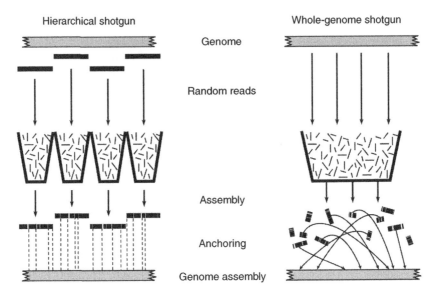

Figure 8.3 Map-based sequencing versus whole-genome shotgun: a scheme of the genome-
sequencing strategies of the Human Genome Project and Celera Genomics
illustrates the different spatialization steps involved.

Source: Waterston *et al.* 2002, reproduced with kind permission of Eric Lander and PNAS.

the flagship of the Human Genome Project. In due course, it became a model for
its organization.

Spanning 100 million biochemical building blocks, the genome of *C. elegans* was
two orders of magnitude larger than any genome that had been sequenced by the
early 1990s. Sulston approached this formidable task by sequencing selected fragments of DNA from the map of the *C. elegans* genome, which he had constructed
several years earlier together with Alan Coulson and Bob Waterston (Coulson *et al.*
1991). This sequencing strategy became known as "map-based sequencing" or
"hierarchical shotgun sequencing." The physical map served him as a convenient
repository of DNA. In addition, sequencing in this step-by-step manner effectively eliminated technical problems associated with whole-genome shotgun
sequencing. As sequence information obtained from the fragments of the physical map was local, the problem of long-range misassembly was eliminated.
Sequencing small portions of the genome also reduced the likelihood of short-
range misassembly. As an early report from Sulston's laboratory justified, this
strategy was feasible with existing technology (Sulston *et al.* 1992). Nonetheless,
other scientists criticized the map-based approach for yielding much non-informative
data from repetitive sequences at an early stage of the project.

While sequencing selected fragments from the physical DNA map simplified
assembly, it also facilitated collaboration. The fragments introduced modularity,
which provided a mechanism for the cooperation of research groups within the

sequencing project. At the Sanger Centre, for instance, separate sequencing teams initially worked on different fragments of DNA. The collaboration with Washington University was organized similarly. As Alan Coulson, then a senior scientist at the Sanger Centre, explained in 1994: "We would not be able to divide the project in a sensible way between us and St Louis, if we didn't have a map to do it in an ordered fashion" (A. Coulson at Sanger Centre Open Day 1994).

At the same time, map-based sequencing began to structure the daily work of the scientists and an increasing number of technicians at the Sanger Centre. Each sequencing team worked on a fragment of DNA from start to finish and routines associated with map-based sequencing began to be differentiated within the team. Later, especially when sequencing human DNA, specialists performed routine steps such as mapping, raw sequence production or checking for errors in the assembly. As the Sanger Centre moved into larger purpose-built facilities on the estate in 1996, the differentiated tasks were also separated spatially. Mapping and sequencing, for instance, were accommodated in separate parts of the new building (Sulston 2000). Increasingly, the Sanger Centre operated like a sequence assembly line, albeit the facade of the country estate was maintained by converting Hinxton Hall into a conference center.

Genome maps not only facilitated cooperation with distant genome-sequencing centers, but connected the Sanger Centre with other researchers studying *C. elegans*. Sequencing the genome in portions at a time meant that finished sequences could be published continuously in an electronic database. This provided other researchers with a continuous stream of intelligible information and helped convince the "worm community" that sequencing at a large, centralized facility was worthwhile (Sulston 2000). The sequences published online were useful in conventional academic research because their genomic provenance had been mapped. Other researchers could draw on the sequences to locate and study the genes of *C. elegans*.

A diagram presented by Alan Coulson at its Open Day in 1994 (Figure 8.4) further illustrates how the Sanger Centre articulated its place in the research community: as a central facility, it would maintain the physical genome map of *C. elegans* and supply the research community with DNA material. It would also curate a database of genomic information on the worm, continuously adding its own sequencing data and genetic mapping information generated by dispersed research groups. In short, the Sanger Centre presented and understood itself as a service provider. As John Sulston commented on this database at the summit of the Human Genome Organisation that year: "I feel it has not been so much a map as a means of genomic communication on *C. elegans*" (J.E. Sulston at the Human Genome Summit 1994). Such talk of cooperation appeased scientists who opposed "big science" and justified the allocation of resources to genome-sequencing facilities like the Sanger Centre. More importantly, it defined the role of the Sanger Centre in the research–industry–technology landscape and ultimately the Human Genome Project.

This historical sketch shows how the physical map of the *C. elegans* genome became the basis of a low-risk sequencing strategy at the Sanger Centre and came

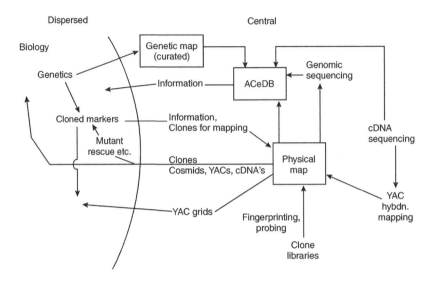

Figure 8.4 Alan Coulson's diagram: presented at Sanger Centre Open Day in 1994, the diagram articulates the place of the Sanger Centre within the traditional biological research community.

Source: A. Coulson, reproduced with his kind permission.

to underpin its collaboration with the Washington University. In addition, this sequencing strategy produced a stream of potentially useful sequence data that was continuously released, convincing many biologists that this mode of organization might be suitable for the Human Genome Project as a whole. Around 1994, laboratories around the world began to form consortia, planning to sequence parts of the human genome in arrangements resembling the collaboration between the Sanger Centre and Washington University. And Tom Caskey, the president of the Human Genome Organisation urged his colleagues that the *C. elegans* sequencing project should be regarded as a "model system of how we should be conducting ourselves" (C.T. Caskey at the Human Genome Summit 1994).

The Institute of Genomic Research

The Institute of Genomic Research was set up in 1992. Its founding director Craig Venter had previously overseen a sequence-based discovery program of human genes at the National Institutes of Health (NIH). When he proposed to expand this program, the extramural research community objected to his plan, which was subsequently dropped by NIH (Cook-Deegan 1994: 311–25). Venter accepted funding from a company instead and set up The Institute of Genomic Research (TIGR). TIGR—Venter pronounced the acronym as "tiger"—was initially embroiled in controversy because it kept sequence information on human genes obtained in contract research in private databases. Later TIGR—its detractors

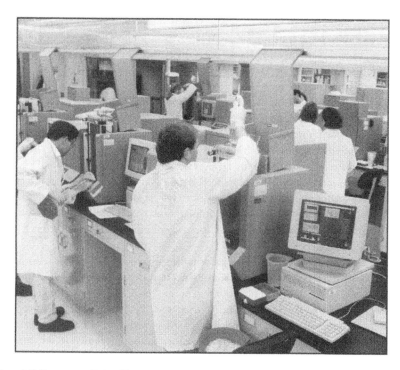

Figure 8.5 Factory or Lab? The interior of The Institute of Genomic Research, a typical genome sequencing centre of the mid-1990s.

Source: Cover of *ASM News*, 62(3), March 1996, reproduced with kind permission of the American Society for Microbiology.

had begun to use the diminutive pronunciation "tigger"—was aggressively promoted as a new model for high-throughput DNA sequencing and analysis (Adams *et al.* 1995). However, the research of this genome-sequencing center (Figure 8.5) evolved into an operation that differed from the Sanger Centre with respect to implicit goals, specialist expertise, and organization. By 1995, TIGR had sequenced the first bacterial genome by means of what came to be known as the "whole-genome shotgun."

The gene discovery program carried out at TIGR in the early 1990s differed from work at the Sanger Centre in two respects: first, by design sequencing experiments carried out at TIGR only yielded sequences corresponding to protein-encoding genes. Indeed, in the early 1990s some scientists argued that this approach should take precedence over map-based sequencing because the coding sequences constituted the major information content of the human genome (see e.g. Adams *et al.* 1991). Second, while the Sanger Centre produced sequences whose genomic provenance had been mapped prior to sequencing, TIGR generated sequence data whose origin in the genome was unknown.

On the one hand, these differences reflected endemic disagreements on the priorities of the Human Genome Project. Was the objective to discover all genes before revealing less promising sequences or to sequence genomes methodically from end to end? On the other hand, these differences in laboratory practice show how locally existing resources were re-invested in future research at both centers: the Sanger Centre drew on a recent accomplishment of its founders, the physical map of the *C. elegans* genome, as a DNA repository. TIGR redeployed computational methods from its early gene discovery program in subsequent genome-sequencing projects. As an early report from TIGR phrased it, computational methods developed to create assemblies from hundreds of thousands sequence shreds "led us to test the hypothesis that segments of DNA several megabases in size, including entire microbial genomes, could be sequenced rapidly, accurately, and cost-effectively by applying a shotgun sequencing strategy to whole genomes" (Fleischmann *et al.* 1995: 496). Features of the laboratory information management at TIGR system were "applicable or easily modified for a genomic sequencing project," too (Fleischmann *et al.* 1995: 497).

The DNA of *Haemophilus influenzae*, the pathogen that causes ear infections, was the first large bacterial genome to be sequenced in this manner at TIGR in 1995 (Fleischmann *et al.* 1995). The entire genome spanning 1.8 million chemical building blocks was fragmented with ultrasound. Fragments of an appropriate size were sequenced from both ends in 23,643 sequencing reactions, generating pairs of sequence data whose separation and orientation relative to one another was known. The data was then assembled with TIGR's specialized software in 30 hours of high-performance computing and yielded 140 sequence assemblies. Gaps between them were then closed by means of targeted experiments. Finally, the correctness of the sequence was verified by comparison with an existing physical map of the genome. "Our finding that the restriction map of the *H. influenzae* Rd genome based on our sequence data is in complete agreement with that previously published further confirms the accuracy of the assembly," assured the report (Fleischmann *et al.* 1995: 508).

This account of a successful "double-barreled" whole-genome shotgun experiment shows that the physical map of the *H. influenzae* genome helped to validate the DNA sequence, but played no visible role in the organization of the genome project itself. Besides, as raw data from shotgun sequencing was unmapped and unintelligible before assembly, TIGR deposited the finished genome sequence in a database, rather than publishing sequences incrementally online for the benefit of the scientific community. Comparison with the Sanger Centre suggests that although the term strategy conjures up the image of a purposeful plan determined by theoretical considerations, the genome-sequencing strategies of both TIGR and the Sanger Centre, in practice, drew on locally existing resources and expertise. Both research centers strove to capture the impetus and align the objectives of the Human Genome Project with their own research agendas. Not surprisingly, Craig Venter asserted that the whole-genome shotgun was fit for carrying out the Human Genome Project: "this strategy has potential to facilitate the sequencing of the human genome," he concluded (Fleischmann *et al.* 1995: 511).

Map-based sequencing versus whole-genome shotgun

In February 1996, representatives from all major genome-sequencing centers convened in Bermuda to evaluate approaches for sequencing the human genome. The delegates—among them, John Sulston and Craig Venter—considered a number of genome-sequencing strategies, but in the end concluded that sequence data had to be generated by the various approaches before they could be evaluated properly. Nevertheless, map-based sequencing became the status quo of the Human Genome Project. The collaborative mechanisms of the *C. elegans* project were extended to human DNA sequencing, albeit with some modifications. The main difference between the *C. elegans* project and the Human Genome Project was that a complete repository of mapped DNA fragments existed from the outset of the *C. elegans* sequencing cooperation, but only a digital set of ordered landmarks was available for the human genome at the time. Consequently, each genome-sequencing center had yet to align suitable pieces of human DNA with this unified map in preparation for sequencing. Although the production of "sequence-ready maps" proved to be the bottleneck in the human DNA sequencing for some time, the leaders of the Human Genome Project felt that this drawback was outweighed by other advantages of map-based sequencing.

Between 1996 and 1998, the digital unified map of the human genome provided mechanisms and metaphors for collaboration and coordination in the Human Genome Project. For example, when the Wellcome Trust announced its intention to fund the sequencing of one-sixth of the genome, the Sanger Centre could register corresponding "sequencing claims" on an internet-based map known as the Human Genome Sequencing Index. The "boundaries" of sequencing claims were defined by means of "unique markers" from the genome map. In addition, the members of the human genome-sequencing consortium endorsed a set of rules, known as "sequencing etiquette," that specified which "regions" each laboratory was entitled to claim based on past sequencing achievements (Bentley *et al.* 1998).

One year after the meeting in Bermuda, Jim Weber from the Marshfield Medical Research Foundation and Gene Myers, a computer scientist at the University of Arizona, made an argument for sequencing the human genome by means of the whole-genome shotgun strategy in the journal *Genome Research* (Weber and Myers 1997). Phil Green, a former member of the *C. elegans* sequencing project, rebutted this proposal in an article published back-to-back in the same volume (Green 1997). He criticized that the whole-genome shotgun offered no mechanisms for upholding the practical values that had held the *C. elegans* collaboration together and were being instituted in the Human Genome Project. In his words, it was "unclear how the project could be distributed among several laboratories" (Green 1997: 416). Green also characterized the whole-genome shotgun regime as "inherently a monolithic approach" (Green 1997: 411), which in terms of organization might be read as a statement of Lewis Mumford's classic thesis that technologies are intrinsically authoritarian or democratic. After all, the whole-genome shotgun was at the time only carried out single-handedly at TIGR.

According to Green, the advantages of map-based sequencing went beyond coordination. The strategy also made room for accommodating differences between laboratories, as well as providing means for isolating problematic regions. In this manner the Human Genome Project could be partitioned between investigators "without forcing them to interact" (Green 1997: 411), and laboratories could explore "alternative sequencing strategies and methods independently without redundancy of effort" (Green 1997: 410). Any unintended replication of sequencing efforts was considered to amount to unwarranted and costly duplication. Besides, the map-based sequencing regime permitted "to isolate problematic regions" (Green 1997: 411). Whole-genome shotgun sequencing, in contrast, offered no mechanism for separating contributions from individual laboratories and for isolating problematic regions: "Problems with data quality in one laboratory would affect all laboratories, because any region of the genome would have shotgun reads generated at all labs," lamented Green (1997: 416).

In making the case for whole-genome shotgun, Weber and Myers negated some of these alleged advantages and left others unanswered. Regarding collaboration, for instance, they suggested, contrary to Green's claim, that laboratories "throughout the world could participate in raw sequence generation" (Weber and Myers 1997: 401), effectively a proposal to decentralize sequence production. Most importantly, they proposed a distinct role for the unified map of the genome. In their sequencing regime the consensus map would function as a "scaffold" for the assembly of the human DNA sequence, not as a framework for the organization of the Human Genome Project. According to Weber and Myers, the task of assembly was to build "scaffolds that span adjacent STSs" (Weber and Myers 1997: 403). Since it was available, information from the unified map of the human genome could assist assembly of the whole-genome shotgun, but would play no role in the organization of the program.

In short, the sequencing strategy of the Human Genome Project constituted a moral and productive economy, whose common ground, so to speak, was the map of the human genome. Reminiscent of the *C. elegans* project, the map underpinned cooperation and coordination, but also came to orchestrate access to resources, accountability for cost and allocation of scientific credit. Although the whole-genome shotgun was organized around the same automatic DNA sequencing machines, its moral, productive, and epistemic standards were different to the extent that Green argued that it was "incompatible" with map-based sequencing (Green 1997: 411). Having explored the mechanisms of collaboration and cooperation, the analysis will now focus on the different products emerging from the two genome-sequencing strategies.

Assumptions and consequences

Throughout the history of the Human Genome Project, the human genome—according to the vision of its leaders—had been envisaged to be a single and coherent domain—24 unbroken sequences of genetic code corresponding to the human chromosomes. The dispute in the pages of *Genome Research* thrived on this assumption: the arguments of the adversaries were framed primarily in terms of feasibility and cost, albeit they also overtly disagreed on the objectives of the Human Genome Project. Between the lines, they also acknowledged that their strategies were based on variant assumptions about the nature of the human genome, on the one hand, and would ultimately manifest different versions of the genome, on the other. Their remarks are indicative of differences beyond superficial sequence similarities and discrepancies.

For Phil Green, the advocate of map-based sequencing, the ultimate aim of the Human Genome Project was to generate an accurate and complete human DNA sequence, which was to serve as a "reference against which human variation can be catalogued" (Green 1997: 410). This aim, he felt, would be best attained by map-based sequencing, which permitted to isolate and carefully sequence problematic regions. For Weber and Myers, in contrast, the true objective of the Human Genome Project was to "sequence all human genes" or, more pragmatically, "to generate as much of the critical sequence information as rapidly as possible and leave clean-up of gaps and problematic regions for future years" (Weber and Myers 1997: 406). Weber and Myers argued that these objectives could be achieved with minimum effort with the whole-genome shotgun strategy, which had emerged from a sequence-based gene discovery program at TIGR.

The opponents also admitted that their approaches would manifest different versions of the human genome. Phil Green, the advocate of map-based sequencing, anticipated that the whole-genome shotgun had "a high probability of failure" by which he implied that the final assembly would contain a very large number of gaps and fail to reconstruct problematic regions faithfully (Green 1997: 411). Jim Weber and Gene Myers conceded that their strategy would by no means produce in a "single unbroken sequence for entire chromosomes," which was perfectly reconcilable with their priorities (Weber and Myers 1997: 404). They found fault with map-based sequencing because rearrangements in DNA fragments required for map-based sequencing could cause artifacts. In their eyes, map-based sequencing was working towards an "arbitrary, mythical goal of 99.99% accuracy of a single, artifactual (in places) and nonrepresentative copy of the genome" (Weber and Myers 1997: 406). Green countered that the artifacts caused by rearrangements could be prevented easily in map-based sequencing.

Falling under the heading of assumptions about the nature of the human genome, Green criticized that, in computer simulation of the whole-genome shotgun, Weber and Myers had assumed that repeated sequences are distributed uniformly throughout the genome. Green asserted this was an "incorrect assumption about the nature of the genome" (Green 1997: 411). Similarly, one might suggest that Green's advocacy of mechanisms for managing problematic regions might have arisen from common assumptions about the nature of the genome. For Green, arguing from the vantage point of map-based sequencing, it was natural to assume that the human genome had "regions," which were bounded by landmarks of the consensus map or embodied in physical DNA fragments. Crucially, by virtue of these regions problematic data could be isolated during the incremental analysis of the human DNA sequence. The spatial metaphors used to manage the Human Genome Project became properties of the human genome itself and a resource for its organization. Weber and Myers, in contrast, articulated no mechanism for dealing with problematic data. They simply proposed, "only a few or possibly even one large informatics group would assay the primary task of sequence assembly" (Weber and Myers 1997: 401). In other words, they proposed not to regionalize the human genome, but to divide the community of genome researchers into those responsible for raw sequence production and those responsible for assembly.

Finally, Green's assertion that the whole-genome shotgun was "inherently monolithic" can be read as a hint at deeper differences. In a whole-genome shotgun regime, the human genome sequence emerged as a whole only after a painstaking process of sequence assembly. In map-based sequencing, in contrast, the genome sequence emerged continuously and incrementally as a mosaic of contributions sequenced and assembled at dispersed genome-sequencing centers. The differences between the genome sequences produced by means of the two strategies might therefore reside beyond the surface of sequence similarities and discrepancies. For example, if problems with sequence data or the assembly algorithms are discovered in the future, what will be the effect on the respective sequences? For a sequence derived by means of whole-genome shotgun one might expect global adjustments, whereas consequences for the product of the map-based regime might well be local. As a generalization about these differences, one might say that the human genome, as envisaged and produced by the Human Genome Project, had the topology of a map. Its large-scale specialization was a result of preliminary physical mapping projects. The human genome produced by means of a whole-genome shotgun might be better conceptualized as an algorithm. The spatialization of sequence data emerged primarily during the assembly of sequence overlaps by means of specialized computer algorithms developed at TIGR.

Celera Genomics

One year after the controversy in *Genome Research*, Craig Venter deserted the collaborative framework of the Human Genome Project to become the director of Celera Genomics, a company intending to sequence the human genome single-handedly by means of the whole-genome shotgun (Venter *et al*. 1998). The leaders of the Human Genome Project perceived the venture announced on May 9 as a threat. Venter's plan amounted to sequencing the human genome significantly faster and cheaper than the Human Genome Project, which had by 1998 merely deciphered 5% of the genome. The defense of the Human Genome Project had four prongs: First, in a series of scientific articles its leaders justified their sequencing strategy (Bentley *et al*. 1998; The Sanger Centre and The Washington University Genome Sequencing Center 1998; Waterston and Sulston 1998). Maynard Olson and Phil Green, for instance, argued that it was "essential to adopt a 'quality-first' credo" on "both scientific and managerial grounds" (Olson and Green 1998). Second, they explained the strategy to the scientific community alleging that it was the only approach that could be "efficiently coordinated to minimize overlap between collaborating groups" (quote from The Sanger Centre and The Washington University Genome Sequencing Center 1998: 1099; Dunham *et al.* 1999) Third, the strategy of the Human Genome Project was revised considerably several months later to allow for the production of a preliminary draft of the human genome. Finally, arguments for the continuation of the Human Genome Project went hand in hand with public criticism of Craig Venter and disapproval of his sequencing strategy. Craig Venter was cast in the role of an unscrupulous profiteer. In plain words, his method was often dismissed as "quick and dirty."

While I can judge neither the character nor the motivation of any of the actors, I suspect that attempting to "shotgun" the human genome might have been a question of personal pride for Craig Venter. Having pushed the limits of shotgun sequencing at TIGR and later seen his method rejected by the scientific community, he may instead have teamed-up with a company willing to fund his endeavor. He may also have felt that inadequate use was being made of the facilities available at TIGR under a map-based sequencing regime. For, TIGR was designed for large-scale shotgun sequencing just as much as shotgun sequencing had originally been designed around the resources available at TIGR. At any rate, Craig Venter must have been well aware that this endeavor would yield a product incongruent with the product of the Human Genome Project.

Only one popular article published on the website of the Lasker Foundation (1998) attempted to compare the sequencing strategies employed by Venter and the Human Genome Project in their own right, concluding that the order of mapping and sequencing was reversed in the two approaches: "In a sense, Venter's approach is to sequence first, map later. In the public project, genes are first mapped to a specific, relatively small region of the genome, then sequenced." The article highlighted that both sequencing strategies drew on existing STS-maps of the human genome: The Human Genome Project used STS-maps prior to sequencing as described earlier. Celera Genomics searched for STS in its primary sequence assemblies, thus establishing their genomic provenance in a secondary specialization step called "anchoring." Despite its methodological focus, this article pitted the two strategies against each other in terms of feasibility.

After the formation of Celera Genomics, debate was polarized in public to suggest that Celera Genomics might keep its sequence data secret and establish a monopoly by patenting the majority of human genes. These motives were cited as the main reasons for the inability of the Human Genome Project to cooperate with Celera Genomics. As the controversy in the pages of *Genome Research* had shown, another important obstacle to cooperation resided in the fact that the whole-genome shotgun provided no proven mechanisms for upholding the practical values that had made the *C. elegans* project work. In public, the lack of these mechanisms was articulated in terms of gene patenting and data release.

Patenting

In presenting his business plan, Craig Venter assured the readers of *Science* that "we do not plan to seek patents on primary human genome sequences" but to identify and patent no more than "100 to 300 novel gene systems from among the thousands of potential targets" for medical interventions (Venter *et al.* 1998: 1541). The leaders of the Human Genome Project, in contrast, portrayed the private venture as contrary to the public interest. "Underlying all this is concern about a land grab—that is we're concerned about massive patenting," commented one scientist (R. Gibbs, quoted in *The Australian* March 31, 1999), and in a sense his remark was accurate: The venture of Celera Genomics amounted to a "land grab" in so far as the human genome was conceptualized as a territory by the Human Genome Project. In proposing the whole-genome shotgun, Venter ignored the sequencing etiquette and all sequencing claims that had already been registered on the Human Genome Sequencing Index. As Venter diagnosed the cause of irritation, "we are changing the rules and that upsets people" (J.C. Venter, quoted in Larkin 1999: 2218).

Data release

Another outcome of the talks in Bermuda had been an agreement to make the human DNA sequence publicly available as it was produced. According to the "Bermuda principles" human sequence data "should be freely available and in the public domain in order to encourage research and development and to maximise its benefit to society" (Smith and Carrano 1996). In practice, this meant that contributing laboratories had to publish assembled sequence data on the Internet within 24 hours, or else faced exclusion from the human genome-sequencing consortium. Another rudiment of the *C. elegans*-sequencing project, this policy aimed to minimize competition among genome-sequencing centers and to deter them from focusing their efforts on lucrative genes. Celera Genomics, in contrast, announced it would release data into the public domain at least every three months and assured that, being an information company, "complete public availability of the sequence data" is "an essential feature of the business plan" (Venter *et al.* 1998: 1541).

Once again, the policy of the Human Genome Project had been justified on epistemic and managerial grounds in 1996. "The finished sequence should be released directly upon completion," argued a scientist of the Sanger Centre, insisting that this was required both to "optimise coordination" and to facilitate "independent checking" (Bentley 1996: 533). More generally, the laboratories of the Human Genome Project planned to implement a dispersed regime of data quality monitoring in 1996: each laboratory sequenced regions of the genome but it was envisaged that the ultimate validation of the data would arise from independent checking by other laboratories. To this end, the network of laboratories even proposed a data exchange exercise, in which the same raw data would be assembled at different laboratories to check for discrepancies. In contrast to the immediate data release policy of the Human Genome Project, the custom at TIGR was to sequence, check, and annotate sequence data single-handedly before publication. Hence, on epistemic grounds, Craig Venter and his senior scientist Mark Adams objected that quality control would be compromised if immediate data release was enforced at all genome-sequencing centers (Adams and Venter 1996).

Moreover, despite the managerial justification for immediate data release, compliance with the Bermuda principles was by no means a necessary condition for coordination, as illustrated by the data release practices of the yeast genome-sequencing project. This genome-sequencing project was carried out by a network of 79 laboratories and coordinated by means of a map of its genome. Delayed data release was an integral part of the collaboration. For example, when the completion of sequencing the yeast genome was announced in March 1996, 20% of it was still withheld from public databases until industrial sponsors and investigating scientists had satisfied their own interests. André Goffeau, the administrator of the yeast genome project, defended delayed data release "on the grounds that the scientists who did the work deserve to be the first to reap some benefits. 'We cannot just give this away,' he said—a view shared by some researchers—especially those from small labs." (Kahn 1996) If truth were told, several of the large genome sequencing centers that participated in the Human Genome Project had earlier contributed in the yeast-sequencing project. Although differences in data release practices caused tensions between the projects, both collaborations hung together as long as laboratories restricted themselves to sequencing regions of the genome within a map-based sequencing regime.

After the formation of Celera Genomics, the leaders of the Human Genome Project suggested that, depending on future business expedients, the company might refuse to make its sequence data available to the public. Francis Collins, director of the National Human Genome Research Institute, argued as much on June 17 at a hearing in the US House of Representatives. The Human Genome Project was "the best insurance that the data are publicly accessible," he said (F.S. Collins in US House of Representatives 1998: 80). Admittedly, in March 2000, an attempt at cooperation between the two camps fell through because of disagreements on data release. However, on this occasion a failure in reconciling the legal and commercial obligations to which each of the data sets were bound caused the downfall of the negotiations, not a categorical refusal by Celera Genomics to publish or combine its data with that of the Human Genome Project (reported by Marshall 2000).

In February 2001, Celera Genomics and the Human Genome Project finally published their preliminary drafts of the human genome simultaneously but separately in *Nature* and *Science*. Celera Genomics made its sequence available to academic and nonprofit researchers around the world. Albeit updates of the sequence will be available to paying customers only, the company has arguably honored its assurance to make the sequence publicly available. Spokesmen of the Human Genome Project continue to criticize the terms that restrict the uses of the sequence data produced by the company. Debates over the conditions of access to the sequence data produced by publicly funded genome-sequencing centers also continue to simmer (reported by Roberts 2002), as does the controversy about the propriety of patenting human genes. Celera Genomics has filed significantly more than 300 provisional patent applications, but maintains that, as anticipated, no more than 300 patents will be filed eventually.

Even after the publication of the two human genome sequences in February 2001, another controversy arose. To save time and money, Celera Genomics had supplemented its own shotgun sequence data with publicly available data from the Human Genome Project. The company had shredded this data, combined it with its own and assembled the combined set. Leaders of the Human Genome Project now charged that the data had not lost its positional information and that the combined assembly had succeeded only because of the inclusion of data produced by the Human Genome Project. In effect, they argued that Celera Genomics had not produced an independent human genome sequence but merely replicated the product of the Human Genome Project. This charge was repeated in a peer-reviewed journal in March 2002 (Waterston *et al.*) and rebutted in the same volume (Myers *et al.*). According to one newspaper report, Craig Venter countered that he had "shredded the data specifically to loose any positional information because he suspected the data had been misassembled in places" (Wade 2001).

Conclusion

Although the human genome has long been conceived of as single natural domain, Celera Genomics and the Human Genome Project have recently produced and will continue to produce different versions of the human genetic code. The Human Genome Project produced its draft as a mosaic of contributions from different laboratories and with the help of long-range maps. The genome produced at Celera Genomics, in contrast, emerged from a whole-genome shotgun experiment. Here the primary spatialization of data emerged during sequence assembly.

In February 2001, both drafts of the human genome had on the order of 100,000 gaps, were riddled by known and unknown inaccuracies and errors, and differed in overall length. As both versions are gradually improved, it remains to be seen, whether the two products turn out to be as "incompatible" as the methods by which they were produced, "complementary," as suggested when leaders of the Human Genome Project attempted to make peace with Celera Genomics in 1999 (Lander 1999), or "hardly comparable" because "different procedures and measures were used to process the data" (Bork and Copley 2001). As the most recent dispute between the Human Genome Project and Celera Genomics demonstrates (Myers *et al.* 2002; Waterston *et al.* 2002), the precise role of genome maps in the production of the two drafts published in February 2001 also continues to be debated.

Even if the two versions of the human genome were to be synthesized, practical mechanisms for doing so will have to be negotiated: one option would be to combine the underlying data for joint analysis, which is arguably what Celera Genomics has already done. However, as the latest in a long series of controversies between the two camps illustrates, such cooperation would require negotiations about the status and value of positional information of map-based data. Another conceivable approach would be to collate the two versions of the human DNA sequence and resolve discrepancies by further experiments. Finally, if all attempts to reconcile the two data bases fail and the sequences remain separate, it remains to be seen whether they will ultimately be used in different ways, one as a reference sequence for basic research and public health, the other to satisfy the information needs of pharmaceutical companies.

Acknowledgments

I am indebted to John Sulston, who made himself available for an interview, and Alan Coulson, who provided a copy of his diagram from the Sanger Centre Open Day in 1994. Don Powell, press officer, has also been very helpful at the Sanger Centre, as has Barbara Skene of the Wellcome Trust. I thank the Studienstiftung des deutschen Volkes and Trinity College, Cambridge, for scholarships for a master's degree during which I undertook most of the research assembled in this chapter. Helen Verran and David Turnbull helped me in making sense of it. Katrina Dean read and criticized several of the preliminary drafts of this chapter.

Notes

There are some aspects of the genomes controversy that I have deliberately not engaged with. In general, my argument has been a historical analysis of the two sequencing strategies as different experimental traditions, or epistemic cultures (Knorr-Cetina 1999), not a detailed analysis of the very latest papers published by the two camps. In the same vein, I have not recounted in detail the modifications of their respective strategies by Celera Genomics and Human Genome Project in the tumultuous years from 1999 until 2002. These are discussed in John Sulston's recently published account of the controversy, as well as in the account of a science writer who was an inside observer at Celera Genomics from 1998 until 2000 (Sulston and Ferry 2002; Shreeve 2004). The details of the different assembly algorithms and quality assurance procedures developed by the Human Genome Project and Celera Genomics are yet to be examined. Finally, between May 1999 and March 2000 the genome of the fruit fly *Drosophila melanogaster* was successfully sequenced by means of a whole-genome shotgun in a formal collaboration of Celera Genomics with the Berkeley *Drosophila* Genome Project (Myers *et al.* 2000). Evidently, mechanisms that enabled the collaboration of the company with academic scientists were established in this project. And in December 2002 a consortium of publicly funded laboratories published a draft sequence of the mouse genome, which was produced by means of a whole-genome shotgun. Evidently, new mechanisms that made it possible to carry out the whole-genome shotgun in a network of dispersed laboratories were established in this project, too (Mouse Genome Sequencing Consortium 2002).

Bibliography

The Australian (March 31, 1999) "Clash over Genome Patents," p. 40.

Aach, J., Bulyk, M.L., Church, G.M., Comander, J., Derti, A., and Shendure, J. (2001) "Computational Comparison of Two Draft Sequences of the Human Genome," *Nature*, 409: 856–9.

Adams, M.D. and Venter, J.C. (1996) "Should Non-Peer-Reviewed Raw DNA Sequence Data Release Be Forced on the Scientific Community?" *Science*, 274: 534–6.

Adams, M.D., Kelly, J.M., Gocayne, J.D., Dubnick, M., Polymeropoulos, M.H., Xiao, H., Merril, C.R., Wu, A., Olde, B., Moreno, R.F., Kerlavage, A.R., McCombie, W.R., and Venter, J.C. (1991) "Complementary DNA Sequencing: Expressed Sequence Tags and Human Genome Project," *Science*, 252: 1651–6.

Adams, M.D., Kerlavage, A.R., Kelly, J.M., Gocayne, J.D., Fields, C., Fraser, C.M., and Venter, J.C. (1995) "A Model for High-Throughput Automated DNA Sequencing and Analysis Core Facilities," *Nature*, 368: 474–5.

Anderson, S. (1981) "Shotgun DNA Sequencing Using Cloned DNase I-Generated Fragments," *Nucleic Acids Research*, 9: 3015–27.

Balmer, B. (1996) "Managing Mapping in the Human Genome Project," *Social Studies of Science*, 26: 531–73.

Bentley, D.R. (1996) "Genomic Sequence Information Should Be Released Immediately and Freely in the Public Domain," *Science*, 274: 533–4.

Bentley, D.R., Pruitt, K.D., Deloukas, P., Schuler, G.D., and Ostell, J. (1998) "Coordination of Human Genome Sequencing via a Consensus Framework Map," *TRENDS in Genetics*, 14: 381–4.

Bork, P. and Copley, R. (2001) "Filling in the Gaps," *Nature*, 409: 818–20.

Cook-Deegan, R. (1994) *The Gene Wars: Science, Politics, and the Human Genome Project*, New York, London: W.W. Norten & Co.

Cooper, N.G. (ed.) (1994) *The Human Genome Project: Deciphering the Blueprint of Heredity*, Mill Valley, California: University Science Books.

Coulson, A., Kozono, Y., Lutterbach, B., Shownkeen, R., Sulston, J.E., and Waterston, R. (1991) "YACs and the *C. elegans* Genome," *Bioessays*, 13: 413–17.

Dunham, I., Shimizu, N., Roe, B.A., Chissoe, S., *et al.* (1999) "The DNA Sequence of Human Chromosome 22," *Nature*, 402: 489–95.

Fleischmann, R.D., Adams, M.D., White, O., Clayton, R.A., Kirkness, E.F., Kerlavage, A.R., Bult, C.J., Tomb, J.-F., Dougherty, B.A., Merrick, J.M., *et al.* (1995) "Whole-Genome Random Sequencing and Assembly of *Haemophilus influenzae* Rd," *Science*, 269: 496–512.

Fletcher, L. and Porter, R. (1997) *A Quest for the Code of Life: Genome Analysis at the Wellcome Trust Genome Campus*, London: The Wellcome Trust.

Goodman, L. (1998) "Random Shotgun Fire," *Genome Research*, 8: 567–8.

Green, E.D. (2001) "Strategies for the Systematic Sequencing of Complex Genomes," *Nature Review Genetics*, 2: 573–83.

Green, P. (1997) "Against a Whole-Genome Shotgun," *Genome Research*, 7: 410–17.

Hilgartner, S. (1995) "The Human Genome Project," in S. Jasanoff, G.E. Markle, J.C. Peterson and T.J. Pinch (eds) *Handbook of Science and Technology Studies*, Thousand Oaks, CA: Sage Publications, 302–315.

Human Genome Summit (January 1994) held at Rice University, Houston, Texas VHS recording held at Sanger Centre Library, Hinxton, U.K.

International Human Genome Sequencing Consortium (2001) "Initial Sequencing and Analysis of the Human Genome," *Nature*, 409: 860–921.

Kahn, P. (1996) "Sequencers Split over Data Release," *Science*, 271: 1798.

Knorr-Cetina, K. (1999) *Epistemic Cultures: How the Sciences Make Knowledge*, Cambridge, MA: Harvard University Press.

Lander, E.S. (1999) "Shared Principles," letter to Celera Genomics (December 28, 1999).

Larkin, M. (1999) "J. Craig Venter: Sequencing Genomes His Way," *The Lancet*, 353: 2218.

Lasker Foundation (1998) "Public, Private Projects at Odds over Research Technique" published September 30 in *Comment*, a publication of Mary Woodard Lasker Charitable Trust. Available at <http://www.laskerfoundation.org/comment/12/comm1.html> (accessed October 5, 2001).

Marshall, E. (2000) "Talks of Public–Private Deal End in Acrimony," *Science*, 287: 1723–5.

Mouse Genome Sequencing Consortium (2002) "Initial Sequencing and Comparative Analysis of the Mouse Genome," *Nature*, 420: 520–62.

Myers, E.W., Sutton, G.G., Delcher, A.L., Dew, I.M., Fasulo, D.P., Flanigan, M.J., Kravitz, S.A., Mobarry, C.M., Reinert, K.H., Remington, K.A., Anson, E.L., Bolanos, R.A., Chou, H.H., Jordan, C.M., Halpern, A.L., Lonardi, S., Beasley, E.M., Brandon, R.C., Chen, L., Dunn, P.J., Lai, Z., Liang, Y., Nusskern, D.R., Zhan, M., Zhang, Q., Zheng, X., Rubin, G.M., Adams, M.D., and Venter, J.C. (2000) "A Whole-Genome Assembly of Drosophila," *Science*, 287: 2196–204.

Myers, E.W., Sutton, G.G., Smith, H.O., Adams, M.D., and Venter, J.C. (2002) "On the Sequencing and Assembly of the Human Genome," *Proceedings of the National Academy of Sciences USA*, 99: 4145–6.

Olson, M. and Green, P. (1998) "A 'Quality-First' Credo for the Human Genome Project," *Genome Research*, 8: 414–15.

Olson, M., Hood, L., Cantor, C.R., and Botstein, D. (1989) "A Common Language for Physical Mapping of the Human Genome," *Science*, 245: 1434–35.

Roberts, L. (2002) "A Tussle over the Rules for DNA Data Sharing," *Science*, 298: 1312–13.

Sanger Centre Open Day (1994) VHS recording, Sanger Centre Library, Hinxton, UK.

Schuler, G.D., Boguski, M.S., Stewart, E.A.D, Stein L., *et al.* (1996) "A Gene Map of the Human Genome," *Science*, 274: 540–6.

Shreeve, J. (2004) *The Genome War*, New York: Knopf.

Smith, D. and Carrano, A. (1996) "International Large-Scale Sequencing Meeting," *Human Genome News*, 7(6). Available at <http://www.ornl.gov/hgmis/publicat/hgn/hgn.html> (accessed February 2003).

Sulston, J.E. (March 19, 2000) interviewed by the author, Sanger Centre, Hinxton, UK.

Sulston, J.E. and Ferry, G. (2002) *The Common Thread*, London: Bantam Press.

Sulston, J.E., Du, Z., Thomas, K., Wilson, R., Hillier, L., Staden, R., Halloran, N., Green, P., Thierry-Mieg, J., Qiu, L., Dear, S., Coulson, A., Craxton, M., Durbin, R., Berks, M., Metzstein, M., Hawkins, T., Ainscough, R., and Waterston, R. (1992) "The *C. elegans* Genome Sequencing Project: A Beginning," *Nature*, 356: 37–41.

The Sanger Centre and The Washington University Genome Sequencing Center (1998) "Toward a Complete Human Genome Sequence," *Genome Research*, 8: 1197–1108.

Turnbull, D. (2000) *Masons, Tricksters and Cartographers: Comparative Studies in the Sociology of Scientific and Indigenous Knowledge*, Australia: Harwood Academic.

US House of Representatives (1998) Hearing of the Subcommittee on Energy and Environment of the Committee on Science, June 17, "The Human Genome Project: How Private Sector Developments Affect the Government Program," Washington, DC: US Government Printing Office.

Venter, J.C., Adams, M.D., Sutton, G.G., Kerlavage, A.R., Smith, H.O., and Hunkapiller, M. (1998) "Shotgun Sequencing of the Human Genome," *Science*, 280: 1540–2.

Venter, J.C., Adams, M.D., Myers, E.W., *et al.* (2001) "The Sequence of the Human Genome," *Science*, 291: 1304–51.

Wade, N. (May 2, 2001) "Genome Feud Heats Up as Academic Team Accuses Rival of Faulty Work," *The New York Times*, p. A15.

Waterston, R. and Sulston, J.E. (1998) "The Human Genome Project: Reaching the Finish Line," *Science*, 282: 53–4.

Waterston, R.H., Lander, E.S., and Sulston, J.E. (2002) "On the Sequencing of the Human Genome," *Proceedings of the National Academy of Sciences USA*, 99: 3712–16.

Weber, J.L. and Myers, E. (1997) "Human Whole-Genome Shotgun Sequencing," *Genome Research*, 7: 401–9.

9 Decoding relations and disease

The Icelandic Biogenetic Project

Gísli Pálsson

This chapter discusses the mapping concepts and strategies employed by deCode Genetics and the Icelandic Biogenetic Project for the purpose of tracing the genetic background of a number of common diseases. I discuss two kinds of mapping involving, respectively, genealogical and genetic data. Each of the two mapping cultures of genealogists and human geneticists, I argue, has its own discourses and trajectories. Collectively, however, the cartographies of genomes and genealogies illustrate the intensification of the medical gaze (Foucault 1973) with the development of biotechnology and the "medicalization of kinship" (Finkler 2000). A series of images, "Birthmarks," by the Icelandic artist Katrín Sigurðardóttir nicely capture some aspects of these developments (see Figure 9.1). Significantly, these images reproduce pre-natal signatures, the artist's own birthmarks, as geographical maps with spatial scales and contours. Also, they represent tiny fragments of the body as islands, echoing the notion of boundedness couched in much of the imagery associated with the Icelandic gene pool and genealogy. At the same time, they underline the seriousness of ongoing Icelandic debates on the representation and commodification of the "national" body and the tension between individual and collective concerns.

The central idea of the Icelandic Biogenetic Project is the assembly of medical records for the Icelandic population. In December 1998, the Icelandic Parliament passed a bill authorizing the construction of such an assembly, the Health Sector Database. The company deCode Genetics, which outlined original plans for the Database, received the license to construct it, in return for a fee. Founded in 1996, it operates in Iceland, although funded by venture capital funds coordinated in the United States. It has grown rapidly, employing about 400 people at the time of writing, a significant figure in a small economy traditionally focused on fishing. The larger Biogenetic Project in which deCode Genetics is involved will, theoretically, allow for the combination, under specific conditions and for specific purposes, of medical records (the Health Sector Database), genetic information, and family histories. Similar projects are now under way in several other contexts, including Australia, Britain, Canada, Estonia, Norway, Singapore, South Africa, Sweden, and Tonga. One of the lessons of Iceland, deCode Genetics, and the Biogenetic Project lies in confronting the complexity of both the mapping procedures involved and the debates and conflicts that have surrounded the emergence

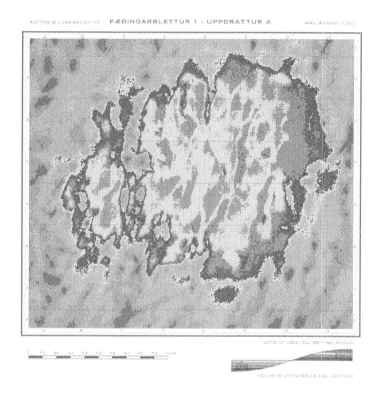

Figure 9.1 "Birthmark 1" by Katrín Sigurðardóttir.

of new modes of connection between cutting-edge research, cultural values, finance, and politics (Pálsson and Rabinow 1999, 2004; Pálsson and Harðardóttir 2002). An important avenue to explore is that of the role of key metaphors. What do people mean when they speak of human genome research in terms of mapping—of charting genealogies and deciphering the "codes" of genetic material— and what are the implications of such a metaphor for human self-understanding? A series of recent publications address the importance of such a question from a variety of perspectives within anthropology, biology, and history of science; see, for example, Rabinow (1999), Kay (2000), Keller (2000).

Recent advances in the mapping of the human genome remind one of the nineteenth-century "scramble" for colonies. Now that most of the habitat of the globe has been charted, documented, and conquered, the West is increasingly turning its attention and interests to the "remotest corners" of living organism, in particular the human genome (Haraway 2000). Perhaps, though, the development of cosmic maps, beyond the Milky Way, represents a more appropriate parallel to genomic maps than projections of Earth and the entire solar system. Both projects have significantly altered the scale and meaning of the cartographic enterprise. Never has "anything," to paraphrase Wilford (2000: 463), encompassed so much and so little space; in the case of cosmic maps, the subject of map-making is infinitely larger than the maps themselves, while for genomic maps the reverse is the case. The cartography of the human genome, with the Columbuses of the New World of modern genetics charting the most minute contours of the human body armed with the perspective of the external observer, underlines the radical separation of nature and society characteristic of modernity. Such a separation was made possible thanks to the development of perspective and positivist science during the Renaissance and the Enlightenment that were to mold every kind of western scholarship. In a relatively brief period nature became a quantifiable, three-dimensional universe appropriated by humans. This "anthropocracy," to use Panofsky's term (1991), represented a radical departure from the enclosed universe of the Aristotelians constituted by the earth and its seven surrounding spheres. A related textual metaphor of the genomic era is that of the "Book of life." Nowadays, it is almost taken for granted that the organic phenomena of genes represent textual codes. This is, in essence, one of the central ideas of "mapping."[1] The mapping cultures of the twentieth century seem to have had some basic ideas in common, including the metaphor of cartographies, a metaphor that echoes the modernist notion of discovery, expansion, and mastery.

The notion of "maps" is obviously used here in a fairly broad sense, for any kind of visual surrogates of spatial relations, including the social space of family histories (genealogies). To think of the tracing of family histories in terms of mapping is not as far-fetched, however, as it may sound. Ingold points out that "the generation of persons within spheres of nurture, and of places in the land, are not separate processes but one and the same" and, therefore, "as Leach has put it, 'kinship is geography'" (Ingold 2000: 149). In fact, representations of kinship have often drawn upon spatial imagery. Medieval European representation of genealogies experimented with various means of visual imagery one of which was that of a flowing stream or a river. An interesting aspect of this experimentation, as we will see, was the theological tension between descending and ascending modes of representing family trees, between "going down" into the soil and "rising up" to the sky. This tension, Klapisch-Zuber points out, "obliges us to ask questions about the connections between language or text and the logic of graphic means of expression which are used to give a visual account of it" (1991: 112; see also Klapisch-Zuber 2000).

The deCode Project: Reykjavík versus Manhattan

Current work at deCode Genetics, independent of the planned Health Sector Database, typically begins with a contract with one or more physicians specializing in a particular disease with a potentially genetic basis. Through a contract with the pharmaceutical company Hoffman-La Roche, deCode Genetics focuses on research on twelve common diseases. In their practice over the years, the physicians have constructed a list of patients with the particular symptoms in question. This list is passed on in an encrypted form to a research team within the company, which, in turn, runs the information provided through its computers, juxtaposing or comparing patients' lists and genealogical records by means of specialized software developed by the company. The aim is to trace the genes responsible for the apparent fact that the disease in question occurs in families. Such an analysis may show, for example, that of an original list of about 1,000 patients, 500 or so cluster in a few families. In the next step, the physicians affiliated with the research team collect blood samples from patients and some of their close relatives for DNA analysis. In the final stage of the research, statisticians evaluate the results of the genetic analysis, attempting to narrow down in probabilistic terms the genes responsible for the disease. In the words of the leader of one of the research teams, "we track the recombination of DNA through each generation, using cross-overs to further localize the location." In practice, this is a highly complex interactive process combining different kinds of mapping, in particular genetic maps indicating genetic distances on a chromosome and more realistic physical mapping. Moreover, strategies of gene hunting are adopted and revised both intuitively in the laboratory or at the computer screen and in formal or informal meetings. These procedures are not, of course, unique to deCode Genetics, not even in the Icelandic context. However, with the Biogenetic Project (see Figure 9.2), with the inclusion of the Health Sector Database and family histories, the power of genetic and epidemiological analyses may grow exponentially, with far larger samples, more generations, and more families. The addition of the national medical records available since 1915 allows for the exploration of a set of new questions on the interaction among a number of variables apart from genetic makeup and genealogical connections, including variables pertaining to lifestyle, physical, and social environments, the use of particular medicine, and degree and kind of hospitalization. The results may be useful, according to the designers of the project and the medical authorities, for pharmaceutical companies and for the medical service, yielding information about potential drugs,

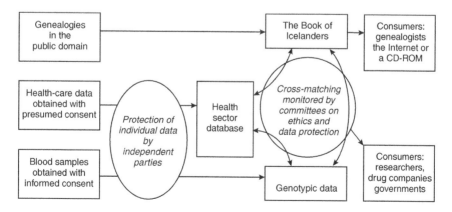

Figure 9.2 The Icelandic Biogenetic Project.

particular genes or proteins, and possible preventive measures in terms of consumption and lifestyle.

Among the many questions raised by Icelandic biotechnology and informatics and their articulation with local and global worlds of commerce is the following: What, if anything, makes the Icelandic gene pool a valuable commodity? Spokespersons for deCode Genetics emphasize that due to its demographic history the Icelandic population may have certain appeals to investors in genetic research. As a result of the successive reduction of the population during times of plagues and famine, they argue, the Icelandic gene pool is fairly homogeneous; branches of the family tree, so to speak, have repeatedly been cut off with a limited number of lineages (*ættir*; "clans" or "families") surviving to the present time. Moreover, geographic isolation has inhibited the replenishment of lost genetic diversity via immigration. Icelanders, however, are no genetic Robinson Crusoes. Throughout its written history, Iceland has regularly been visited by slaves, pirates, fishermen, and travelers. Recent research in biological anthropology (see Helgason *et al.* 2000; Helgason 2001), based on DNA analysis and demographic studies, indeed indicates that while Icelanders are more homogeneous than most other European populations they nevertheless contain a surprising amount of genetic diversity for a small island population.

Whatever the genetic and historical facts of the Icelandic gene pool, the Biogenetic Project has been the center of a controversy. Opposition to the Project seems motivated by several concerns, including potential infringements of personal autonomy and privacy, the ethics of consent, and the threat of biopiracy. Much of the heat of the opposition seems driven by an apparent sense among physicians of rather suddenly lacking authority in the biomedical domain, of losing dominion over public information largely constructed and controlled by them in the past. The Icelandic opposition, thus, above all reveals concerns about potentially dramatic changes in the practices and structures of Icelandic bioscience and medicine. Some academics have alleged that the restrictions of access to information and resources implied in the privileged contract of deCode Genetics with the Icelandic state will inevitably result in the stagnation of bioscience. Thus, the sub-text of some of the debates centers on where cutting-edge research on the Icelandic human genome occurs today and where it should be located in the future. Understandably, proponents of the project represent a more complex professional and social spectrum than the opponents, given the fact that they by far outnumber the critics. Those who have contributed op-ed pieces in support of the project emphasize concerns rather different from those of the opponents. DeCode Genetics and the Biogenetic Project, they argue, will make possible much more biomedical research in Iceland, creating many new positions for scientists and laboratory assistants both within the company and at the University of Iceland. DeCode Genetics, it is often pointed out, has attracted much investment from abroad and created numerous jobs for Icelandic scientists, many of whom in the absence of deCode Genetics would have been forced to seek employment abroad.

Apparently, the deCode Genetics project largely appeals to scientists, investors, shareholders, and the public imagination on the grounds that the Icelandic population has been small (275,000 today), relatively homogeneous, comparatively isolated, and passionate about keeping medical and genealogical records. An important asset for the Biogenetic Project is the availability of relatively detailed and complete family histories. Frisk Software in Reykjavík originally began the construction of a genealogical database on Icelanders early in the 1990s, starting with three censuses (1703, 1801, and 1910) that covered the whole country at sufficiently different points in time, to minimize overlap, as well as the up-to-date national records (*þjóðskrá*). Later on, deCode Genetics signed an agreement with Frisk Software to speed up the construction of the database, now called the "Book of Icelanders," by adding information from a variety of available sources. The database now comprises twelve censuses taken from 1703 to 1930. When completed or "published," the genealogical database will be issued to deCode Genetics (Pálsson 2002). No names will be included, only numbers or IDs that allow for the combination of different datasets on a limited basis for particular purposes. A complex process of encryption, surveillance, and monitoring is designed to prevent illegitimate use of the data. Such a database, it is argued, combined with genetic and medical data, provides an invaluable historical dimension to the search for the presumed genetic causes of common diseases.

It has taken about four years for a team of researchers and computer programmers to compile all of the information contained in the Book of Icelanders and to design the necessary programs for displaying and analyzing it. Approximately 650,000 people, less than a half of the total number of people born in Iceland since the Norse settlement, are recorded in the genealogical database. The whole point, of course, is not simply to record individuals but rather to be able to connect them to each other. The "connectivity index", the proportion of documented genealogical links between the individuals and their parents in the database, is close to 85%. Not only are there empty spaces in available records, there are lots of errors, too. Adoptions and wrong parenting pose a particular problem. Sometimes, families have "purified" their records, possibly to prevent disclosing information about teenage mothers or to avoid an image of inbreeding. Finally, at times the makers of the database are faced with evaluating conflicting pieces of information. Skúlason, the chief architect of the genealogical database, has likened the task to "working out a puzzle the size of a soccer field, with half of the pieces missing and the rest randomly scattered," a gigantic task indeed.

The founding premise of deCode Genetics is that the company is uniquely placed to achieve major breakthroughs in identifying the genes involved in the most common diseases of the advanced industrial world. DeCode's vision, or program, of how to exploit available resources to the fullest is spelled out by Kári Stefánsson (Director) and Jeff Gulcher (Vice President, Research and Development). While the nuclear family has proved to be a useful unit in the study of monogenic disorders, they argue, for complex or "polygenic" disorders that are "sporadic," skipping generations, "isolated populations with strong founder effects" are essential (Gulcher and Stefánsson 1998: 523). For the deCode team, the royal road to identifying the underlying cause of pathology is through linkage analysis, the aim of which is

> ...to determine whether there exist pieces of the genome that are passed down through each of several families with multiple patients in a pattern that is consistent with a particular inheritance model and that is unlikely to occur by chance alone. The most important asset of linkage analysis is its ability to screen the entire genome with a framework set of markers; this makes it a hypothesis-independent and cost-effective approach to finding disease genes.
>
> (Gulcher *et al.* 2001: 264)

For the spokespersons of deCode's project, then, a relatively homogeneous population with good genealogical as well as medical records (for precise phenotypic identification) is the ideal experimental site for linkage, linkage being, in their view, the ideal method of analysis.

Not only have critics of the Biogenetic Project raised concerns about its social implications, there are doubts as well about its scientific merits and rewards. Both the application of linkage analysis for the purpose of understanding complex diseases and the choice of relatively homogeneous and isolated populations remain contested. Thus, Risch (2000) suggests that while multiplex families and linkage studies will have a role to play on the future agenda, "instead of family-based controls, unrelated controls will emerge as a more powerful and efficient approach" and, moreover, sampling families "of varying ethnicity will... be advantageous from the perspective of enhancing evidence of causality as well as identifying genetic and/or environmental modifying factors" (Risch 2000: 855). Similar arguments were raised in a news feature in *Nature* (2000 (406): 340–2) under the title "Manhattan versus Reykjavik."

Modern biotechnology is shrouded in "genohype" (see Fleising 2001). While it is often difficult to disentangle the real, the virtual, and the rhetoric, theoretical and methodological issues such as these should, of course, be evaluated in terms of actual results. Some of the results of the deCode project may be impressive, including the mapping of Parkinson's disease (Gulcher *et al.* 2001: 266–7) and the demonstration of a strong familial component to longevity (Gudmundsson *et al.* 2000), but the significant breakthroughs repeatedly promised by the company and its main corporate financier, Hoffman-La Roche, have failed to appear. One possibility is that the linkage paradigm is in crisis. Another possibility is that success is just around the corner.

Mapping relatedness

Given the centrality of familial relations in research projects such as those of deCode Genetic, it is pertinent to ask: What does relatedness involve and how is it understood and represented? With the works of Schneider (1984), Strathern

(1992) and some others, the "classic" anthropological issue of kinship, of mapping relations through descent, was denaturalized and destabilized. Kinship, it was argued, is not necessarily "the same thing" in different contexts, to be routinely disentangled with the formal tools of the "genealogical method." Moreover, kinship, it seems, or rather *relatedness*, that is bonding deemed to be significant by the people involved (Carsten 2000), is best regarded as a complex *processual* thing, not a fixed, abstract geometry of social relationships. Relatedness, in sum, is continually "under construction." To some extent, recent developments in biotechnology have provoked such a conclusion. With artificial reproductive models, gene therapy, and genetic engineering, relatedness becomes not simply a matter of genealogical history but also a matter of consumer choice in a quite literal sense.

While, however, new reproductive technologies have challenged the structures of kinship, they should, perhaps, not be overrated. As Laqueur points out in his survey and evaluation of the "strangeness" of connectedness in the age of sperm banks, ovum brokerages, surrogate motherhood, and similar reproductive techniques and practices:

> This strangeness . . . is not born of technology. It is at the root of kinship: heterodox ways of making families—high, low, or no technology—only expose it for what it is, as if our own, Western, structures were exposed to the scrutiny anthropologists more usually reserve for other people.
>
> (Laqueur 2001: 79)

Laqueur concludes that "so far blood and flesh, mother and father, and sire and dam are as clear or as muddled as they have ever been" (Laqueur 2001: 94). Somewhat paradoxically, while kinship analysis is in a state of flux, molecular biology has given genetic links a renewed, supreme status where there is little room for culture and social construction. And kinship diagrams are back on the scene, "deeper" and more elegant than before.

It may be tempting to think of genetic and genealogical models as being "pure" descriptions of relations and histories, untainted by the contexts that produce them. This is clearly not the case, however. The algebra of the kinship diagram, as Bouquet argues, has "its own historicity, making it an anything but neutral instrument" (2000: 187). This historicity is manifested in a family resemblance of family trees, of "pedigree thinking," in a variety of disciplines. Darwin attempted to locate *homo sapiens* in the genealogy of life, a project, as Beer shows, that reflected the ideals of his society and literature: "He sought the restoration of familial ties, the discovery of a lost inheritance, the restitution of pious memory, a genealogical enterprise" (1986: 221). Just as genealogical diagrams were essential for tracking the paths of human nobility, Darwin reasoned, they were indispensable for naturalists interested in describing relations among other living things.[2]

Klapisch-Zuber (1991) provides an intriguing analysis of the genealogy and politics of lineage imagery in medieval European family histories. Medieval obsession with genealogical details and their visual representation, informed by the politics of inheritance and succession among European elites, is nicely demonstrated by some family diagrams drawn on rolls of parchment; avoiding the interruption of a new page, some of them were no less than ten meters in length. Over time European lineage imagery was subject to much experimenting. One powerful image was that of the flowering tree, which underlined the joyful proliferation of the lineage, drawing its vital energy from the earth and stretching into the divine light in the heavens. Such a horticultural metaphor appeared in various forms from at least the eleventh century onward, but it was only in the fifteenth century that it acquired its canonical imagery, with a founding ancestor in the trunk of a tree and his descendants scattered above among its branches. An alternative imagery turned the tree upside-down, with the trunk in the sky. Such an image is reflected in Latin terminology of filiation (*descendentes, progenies,* etc.). A lower position in such a scheme not only indicated a chronologically later moment, it also suggested deterioration or demotion, a departure from the honored distant past. The descending imagery both documented the continuity of the lineage, the direct line between the present and the past, and the humble status of contemporary humans underlined in eschatological, Christian schemes.

Despite its joyfulness and its success in representing and reinforcing kinship, the ascending metaphor of the flowering tree was seen to be theologically problematic. Given the need to project the past in glorious terms, the image of the growing tree, with the ancestors (and the gods) in the soil and degenerated contemporaries in the heavens, was bound to be met with resistance if not disgust. The tree metaphor, in fact, was riddled with tensions. Not only was there tension between the competing images of ascending and descending, the ups and downs of kinship, there was also tension between the languages of roots and branches. One way of resolving the tension between the ascending image of the tree and the linguistic usage of "descent" was to place the roots of the family at the top of the image. And yet, such an image suggested multiple origins, not a single divine source, with the roots spreading outwards in different directions. Despite the theological complications and the botanical absurdity of the image of a tree that extends its roots into the sky, the metaphor not only survived but flowered into the present.

The medieval imagery of the tree was not exclusively focused on "blood" connections. Interestingly, sixteenth-century genealogical trees showing the kings of Navarre in Portugal included their wives on discrete branches, mixing blood relationships and marital alliances in the same imageries. Nor should the tree imagery itself be taken for granted. In fact, as Klapisch-Zuber points out (1991: 110), it is "quite possible to adopt other graphic systems, simpler but equally effective, for visually presenting the obsessive repetition of genealogical descriptions." Rival metaphors, including those of the human body and the house, appeared from time to time in artistic representations of relatedness.

Medieval Iceland had its own projects of mapping relations and identities, embedded in church records and population registers (Grétarsdóttir Bender 2002). Frequently, these documents offered in passing brief notes or commentaries about particular personalities, usually in the past tense: "He was intelligent and knowledgeable about many things.... He had a tumor the size of a ptarmigan egg on one of his eyelids and became blind on that eye" (a man born in 1802). Often there are comments about a person's character and social history: "He lost his priesthood due to premature intercourse (*offljóta samsængun*) with his wife" (a man born in 1699). Many of the comments indicate concerns with inbreeding, incest, and adultery: "She had four children with her brother and killed them all, except the youngest one" (a woman born in 1799). Sometimes the records indicate a folk theory of the continuity of familial characteristics, the passing of individual traits from one generation to another: "Among his descendants, certain family characteristics have persisted, including tremendous energy, endurance and dexterity" (a man born in 1755).

The idea that a person's character, capabilities and dispositions are partly determined by what one would now translate as "inheritance" appears early on in some of the Icelandic sagas. Thus *Njál's saga* (*Íslendinga sögur og þættir* 1987, ch. 42), one of the major Family Sagas, asserts that "nurture accounts for one quarter" of a person's dispositions (*fjórðungi bregður til fósturs*).[3]

Osteoarthritis: a case study

To illustrate the deCode approach to the exploration of the role of familial relations for explaining differential occurrence of common diseases, it is useful to focus on the team studying osteoarthritis (hereafter "OA"), one of the most common diseases of humans often affecting joints in fingers, knees and hips.[4] The methods used are also discussed in some of the publications derived from the deCode project (see, e.g. Ingvarsson *et al.* 2000; Stefánsson *et al.* 2003). The OA-team was one of the more established teams within the company, with several permanent members, mostly biologists and technical laboratory assistants, collaborating closely with statisticians and physicians specializing in OA. Their project started with the initiative of the physicians. Then the pharmaceutical company Hoffman-La Roche arrived on the scene, and a contract was signed with the clinical collaborators, focusing on two particular phenotypes of the disease, OA of the fingers and hips. Later on, the study of OA of the knees was also incorporated.

Often, it has been assumed that OA "simply" comes with age and drudgery. Indeed, the Icelandic term for osteoarthritis, *slitgigt*, refers to the kind of arthritis that develops as people become "worn down" during the life course. When the deCode project started, it was assumed that there was some underlying genetic factor, since siblings were known to have a higher risk than others of hip replacement. However, no one had successfully "cracked" a complex disease like OA or, in other words, identified the genetic factors involved in the phenotype. The identification of families with the right phenotypes, obviously the critical starting point in work of this kind, is somewhat problematic due to the nature of available sources. S.E. Stefánsson explains: "It's not easy to find well-defined families. One of the problems with past records is that diagnoses often were poor. People didn't know the difference between rheumatoid arthritis and osteoarthritis... By using the 'Book of Icelanders,'" he suggests, his team is able "to show that OA is inherited, there is a founder effect. Simply by going back one generation after another. Patient groups have fewer founders than others. They are significantly different from control groups."

Having established a familial connection, the OA-team has set out to locate the genetic factors involved. S.E. Stefánsson elaborates on the relative advantage of the deCode Genetics team thanks to the "deep" genealogy of the Book of Icelanders: "Most other groups are looking at sib pairs. They have less resolution than we have for linkage analysis as they don't have the genealogies. We only need to know how people are related."

By running their encrypted patient lists (of people diagnosed with OA) against the encrypted version of the Book of Icelanders, the OA-team explores "how people are related." Knowing the genealogical relationships, establishing meiotic distances (the number of links separating any two persons in a pedigree) among the patients, the researchers seek to confirm and narrow down candidate regions, that is, regions with genes whose protein products are assumed to affect the disease.

Figure 9.3, which was drawn by the osteoarthritis team, explores genealogical relationships among a group of patients, in this case two pairs of sisters severely affected by osteoarthritis. The aim, of course, is to trace the genes responsible for the fact that osteoarthritis tends to occur in certain families. The numbers in Figure 9.3 show markers and alleles (alleles being the alternative forms of a gene or DNA sequence at a specific chromosomal location). Thus, the sister-pair on the left have allele 8 in common, possibly from one of the grandparents.

The "Allegro" software developed by deCode Genetics identifies common alleles and calculates lod-scores (z), the likelihood of genetic linkage between loci (unique chromosomal locations defining the position of individual genes or DNA sequences). The OA-team start by genotyping the DNA, using around 1,000 markers, which are dispersed throughout the genome. The alleles from the genotypes are then used to calculate the sharing of the genetic material between related patients. The results, displayed as lod-scores, guide the team to the regions in the chromosome where they expect to find mutations. The findings from the gene hunt are passed on to the company's statisticians who evaluate every now and then how the search is going.

While S.E. Stefánsson is confident about his project, claiming that so far he has "the highest lod-score reported," he emphasizes that he is looking for more than a few genes: "There are many mutations throughout the genome. And there are a number of mutations behind each joint [fingers, hips, and knees], causing the disease. We suspect there are five to ten genes behind each joint."

When pressed about genetic determinism, interaction among genes, and the role of environmental factors, S.E. Stefánsson maintains that while OA is "clearly genetic" since it runs in families, it is a complex disease: "Osteoartritis is probably an interaction between genes. The environment probably doesn't matter much." Sometimes, the team has "genotyped hundreds and hundreds of samples without any lod-scores; everything is just flat." One of the problems

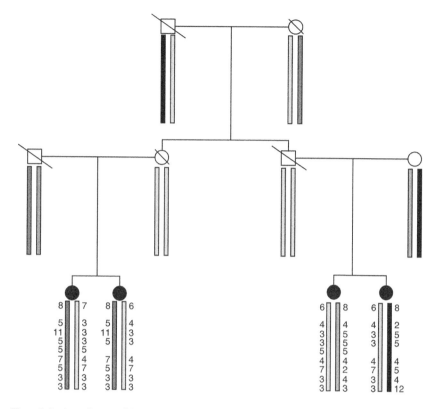

Figure 9.3 A pedigree with two pairs of sisters affected by OA of the fingers.

is that of perceptual bias:

> The problem when you have a candidate gene is that you tend to shift your attention towards it. It's not unbiased. This might have been distracting me ... I spent months for nothing. I was absolutely convinced. We are getting more and more convinced now of a perceptual bias. Everyone in the business does this. You are literally stuck.

Occasionally the statistical results run counter to the expectations of team members who speak of statisticians "slaughtering" their lod-scores: "In November, the statisticians slaughtered our lod-scores. We have a new search now, on slightly different grounds. We need a score of 3.7. I am still confident. I have dense markers."

The hunting for genes is a tiresome job with frustrations and disappointment as well as excitement and euphoria:

> We mapped the chromosome region by region, rebuilt it. Then we narrowed down even further, sequencing base-by-base. There are two candidate genes

left on Chromosome A (a pseudonym). I think I have narrowed-down to a region. I keep on trying until something works, going by hunches. The first step is to use the lod-scores to guide us into the chromosome region. Then we add more markers to the region. If we are lucky, we can see the same markers in unrelated families with the same symptoms. Bingo!

After the complete of the fieldwork on which this chapter is based, the OA-team published an important article about its results. That article reported mapping hand OA to three prominent locations on chromosomes 2, 3, and 4 (Stefánsson *et al.* 2003). Also, it describes a mutation in MATN 3, the gene that encodes matrilin-3 (602109), with an association in some patients diagnosed with hand OA.

To understand the work of the OA-team, it is important to consider its social context. There is extensive competition between companies in the kind of work that the OA-team does and, obviously, it affects the intensity of the work as well as the way in which results are presented: "We could easily have published much more on various chromosomes...but we don't know how much lead we have. Other companies are working on this too."

Also, there is competition between teams within deCode Genetics. While project leaders internal to the company sometimes lean on each other, circumnavigating around common problems, there is competition between leaders and their teams for access to labs and machinery where one team slows others down. Moreover, there are interpersonal and organizational problems of coordination.

Figuring kin

The fragmentary descriptions and glimpses from discussions with the OA-team presented earlier provide some indication about how they represent their reasoning about disease and relatedness. Maps and other forms of visual representation are not only useful tools for research purposes; they are worthy of attention in their own right, from the perspective of the "ethnography of viewing" (Orlove 1991), constituting and reflecting historical activities. And this applies to both genetic maps and the relational maps used to represent genealogies. DeCode Genetics and its scientists and genealogists produce several kinds of genealogical trees. First, a common imagery projects genealogies as descending or going down; one example is the figure with the OA patients presented in Figure 9.3.

Another imagery shows genealogies as ascending into the past, or collapsing into the present, depending on the vantage point. Here, the main mission is to trace "ego's" roots as far as possible, focusing on direct lines between the present and the past rather than the flowering of the whole lineage. Figure 9.4 shows a family tree produced by the Book of Icelanders.

Finally, one kind of imagery presents pedigrees as neither ascending nor descending but as circular outward progressions; such an imagery is evident in the informal logo of deCode Genetics, the asthma-pedigree of Figure 9.5, reproduced on t-shirts and other publicity material (a similar image is also reproduced

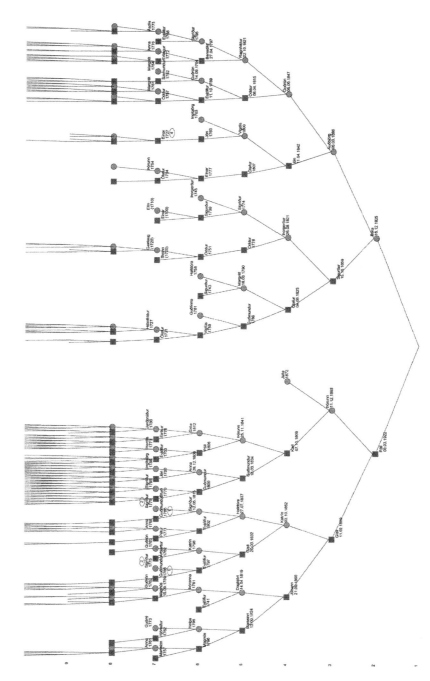

Figure 9.4 A family tree from the "Book of Icelanders."

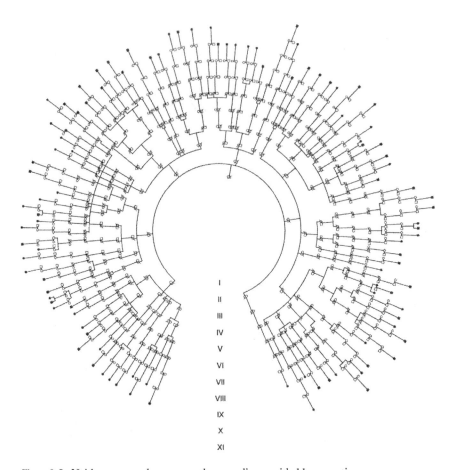

I
II
III
IV
V
VI
VII
VIII
IX
X
XI

Figure 9.5 Neither up nor down: an asthma-pedigree with 11 generations.

in some publications; see Gudmundsson *et al.* 2000: 746). Again, however, the tree imagery seems to be close by. In this case, the progression of time through the generations is represented by an image resembling a cross-section of a trunk of a tree. Just as natural history is engraved in the growth rings of the tree, genealogical history is encoded in the genealogical circle of the Icelandic family tree.

The history of the graphical representation of kinship in medieval Europe indicates that family trees are all but innocent. What could the figurative modes of Icelandic kinship possibly signify? Perhaps, given the variety of representation of genealogies in the deCode Genetics/Frisk Software project, one should not read too much into the differences between these figurative modes in terms of the graphics and layout of pedigrees. To some extent, the designers are "simply" motivated by straightforward visual concerns and the practical constraints of their tools, programs and computer screens, much like medieval genealogists were motivated by the constraints of vellum, paper, and rolls of parchment. On the other hand, the different kinds of images obviously represent particular aspects of the deCode Genetics/Frisk Software genealogical project: one represents the search for mutant genes passed "down" through the generations (Figure 9.3), another reflects the ego-centric search for distant ancestors suggested by the Book of Icelanders (Figure 9.4), and still another underlines the rhetoric of the common roots and interests of the enclosed Icelandic "circle" (Figure 9.5). Apart from its visual appeal, the circular image of Figure 9.5 underlined in the public-relations material of deCode Genetics may serve, consciously or not, to foster a sense of unity and belonging. The reference to the "circle" of the community, the "Book" of Icelanders, and the continuity with the Viking past may, thus, indirectly suggest a rather narrow genetic notion of citizenship. Iceland, in particular Reykjavík the capital city, is rapidly becoming a multicultural community with immigrants from different parts of the world. To some extent, the genetic notion of citizenship implicated by the current database project competes with other notions of citizenship, which have increasingly been gaining force, emphasizing human rights and the empowerment of ethnic groups marginalized in the past.

Trees and plants, it seems, "make perfect natural models for genealogical connections" (Rival 1998: 11). There are countless examples in the ethnographic literature of the metaphoric association of trees and kinship relationships in different times and places. The frequent metaphor of the tree, it may be argued, need not be surprising. Thus, Atran suggests (1990: 35) that as a life form trees may be "phenomenally compelling" to humans generally due to their ecological and historical role. Such a claim, however, seems to violate much ethnography. For Inuit groups in the barren Arctic, for instance, the idea of trees is probably radically different from, say, the trees experienced by inhabitants of South American rain forests (Fernandez 1998: 81). While the life-form of trees may not be evolutionarily significant for human cognition, trees may, as Rival suggests, "be potent symbols of vitality precisely because their status as living organisms is so uncertain" (1998: 23); trees die slowly, unlike animals they do not seem to have a "natural" life span, and, finally, trees may be intensely alive and half-dead at the same time, composed of a mixture of dead and live tissue. Much like, perhaps, family trees.

How and why has the metaphor of the tree assumed the status it has in Western thinking as an organizing imagery? Klapisch-Zuber (1991) leads one to believe that the preoccupation with the tree metaphor has much to do with agrarian discourses on domination and succession. Deleuze and Guattari (1988) seem to agree, suggesting another metaphor, namely the one of the rhizome, a decentered cluster of interlaced threads where everything is potentially interconnected with everything else. Whereas the tree "always comes back 'to the same'," the rhizome has "multiple entryways" (1988: 12).[5] Should a model of relatedness, one may ask, only allow for a once-and-for-all transmission of "natural" substance from one generation to another? Much ethnography suggests different models (see Carsten 2000). In some societies, the transmission of substance is regarded as an ongoing process that includes breast-feeding, a process that may take several years after the moment of birth. More importantly, in some contexts kinship connection have little to do with shared substance. The Iñupiat of Alaska represent a particularly interesting case. Here, personhood is constantly being transformed through practice, through adoption, labor and, above all, naming; personal essence is "largely influenced by names rather than inherited through shared substance" (Bodenhorn 2000: 140).

For biomedical and pharmaceutical companies, the Icelandic gene pool is obviously valuable capital precisely because of the presumed direct genealogical link of "real" Icelanders, through a few lineages, with the medieval past. I have discussed two kinds of mapping interrelated in the deCode Genetics project, involving genealogical and genetic data. While each of the two mapping cultures of genealogists and human geneticists has its own discourses and trajectories, they share, by definition, the notion that relatedness is both determined at the point of conception and a function of the number of "steps" separating any two persons, a notion captured by the "genealogical" image of the tree. The powerful discourse on the "book of life" established during the last century (Kay 2000) and the recent achievements of genetics and biotechnology have fostered a renewed interest in family histories and kinship diagrams. As we have seen, however, family trees are never innocent phenomena in the sense that they have a social life of their own, a biography informed by the contours of the cultural landscapes to which they belong. Genealogical models of trees embody the pragmatics and the social contexts of those who make them, but they also (and this follows from their cultural baggage) tend to inform the contexts where they are applied. As Haraway puts it, "map making is world making" (2000: 112). Thus, the Biogenetic Project and the imagery and rhetoric associated with it may reinforce a rather narrow notion of "Icelandicness," a notion underlined in the nationalist discourse of the last century.[6] Not only is the notion of the blueprint of life surrounded by a persistent and growing uneasiness as it remains unclear what a "gene" is (Keller 2000; Rheinberger 2000), also the inclusion of genealogies in biogenetic projects is a contested issue, partly because their graphic representation invites controversial notions of identity and citizenship. Maps, whether genealogical or genetic, may look deceptively simple. However, a critical engagement with them is an important element of the recent "spatial turn" (Cosgrove 1999: 7) in the sciences and the humanities.

Acknowledgments

Parts of this chapter appeared in Pálsson (2002) and Pálsson and Harðardóttir (2002). The research on which it is based has been supported by several funds, including the Nordic Committee for Social Science Research (NOS-S), the Icelandic Science Fund, and the University of Iceland. I thank Kári Stefánsson, Director of deCode Genetics, and the OA team at deCode Genetics, in particular Stefán Einar Stefánsson, the leader of the team, for their cooperation. Also, I am grateful to James Griesemer (University of California at Davis), Linda F. Hogle (Stanford University), Benjamin S. Orlove (University of California at Davis), Stacy Leigh Pigg (Simon Fraser University), Paul Rabinow (University of California at Berkeley), and the editors of this volume for useful comments and suggestions.

Notes

1 In contrast, during the eighteenth century some linguists seriously entertained the idea of *language* as being literally a natural organism. Thus, Franz Bobb argued that languages must be regarded as "organic bodies, formed in accordance with definite laws; bearing within themselves *an internal principle of life*" (cited in Newmeyer 1986: 23, emphasis added).

2 Pedigree thinking, in fact, is a common Western theme that reveals itself not only in the formal diagrams of kinship studies and evolutionary biology, but also in philology and the "Ethnologia" or *Völkerkunde* of museum collections. The genealogical view of the world outlined the common roots and relatedness of different groups and nations, providing the classificatory rationale for ethnographic collections and museum cabinets. Linguistics and biology presented a purified vision of languages and organisms as timeless artifacts with parallel roots and histories. In anthropology, the genealogical method had important implications for both theory and empirical approach: it "fixed birth as the defining moment of kinship, and fixed the instruments for its recording accordingly" (Bouquet 2000: 187).

3 The significance of genealogies during the saga age does not suggest that Icelanders were preoccupied with the structural relations of kinship. In fact, friendship was a powerful social relationship, judging from the accounts of the Family Sagas (see Durrenberger and Pálsson 1999).

4 I have been engaged in ethnographic research within deCode Genetics, focusing on a research team working on OA. Fieldwork, conducted on and off during year 2000, focused on participating in team meetings and informal coffee-room discussions as well as interviewing all team members, some of them repeatedly. Also, several other people were interviewed, including statisticians, and those in charge of the "Book of Icelanders." Among the questions discussed during many of the interviews were: How does one become a skilful gene hunter? How do team members reason about genes, mapping, the discovery of genes, and the articulation of genes and environment for the explanation of differential occurrence of OA.

5 The claim that tree metaphors are inherently Western and hierarchical is a bit overdrawn. The dualism of Deleuze and Guattari that radically separates the arborescent and patriarchal West, on the one hand, and the democratic and rhizomic South or East, on the other hand, indeed has its critics (Fernandez 1998). The persistent metaphor of the tree, which is at the core of the Western genealogical model, in fact can take several forms, depending, of course, on how one regards trees. After all, as Deleuze and Guattari admit (1988: 17), "trees may correspond to the rhizome, or they may burgeon into a rhizome."

6 In fact, the Biogenetic Project was foreshadowed by another grand plan in the 1960s to centralize a variety of information on Icelanders, a plan at the University of Iceland that drew upon concerns among local scholars since the 1930s with the purity of the "Nordic race" (Pálsson 1995).

Bibliography

Atran, S. (1990) *Cognitive Foundations of Natural History*, Cambridge: Cambridge University Press.

Beer, G. (1986) "'The Face of Nature': Anthropometric Elements in the Language of *The Origin of Species*," in L.J. Jordanova (ed.) *The Languages of Nature: Critical Essays on Science and Literature*, London: Free Association Books.

Bodenhorn, B. (2000) "'He Used to Be My Relative': Exploring the Bases of Relatedness among the Iñupiat of Northern Alaska," in J. Carsten (ed.) *Cultures of Relatedness: New Approaches to the Study of Kinship*, Cambridge: Cambridge University Press.

Bouquet, M. (2000) "Figures of Relations: Reconnecting Kinship Studies and Museum Collections," in J. Carsten (ed.) *Cultures of Relatedness: New Approaches to the Study of Kinship*, Cambridge: Cambridge University Press.

Carsten, J. (ed.) (2000) *Cultures of Relatedness: New Approaches to the Study of Kinship*, Cambridge: Cambridge University Press.

Cosgrove, D. (1999) "Introduction: Mapping Meaning," in D. Cosgrove (ed.) *Mappings*, London: Reaction Books.

Deleuze, G. and Guattari, F. (1988) *A Thousand Plateaus: Capitalism and Schizophrenia*, London: Athlone Press.

Durrenberger, E.P. and Pálsson, G. (1999) "The Importance of Friendship in the Absence of States, According to the Icelandic Sagas," in S. Bell and S. Coleman (eds) *The Anthropology of Friendship: Beyond the Community of Kinship*, Oxford: Berg Publishers.

Fernandez, J.W. (1998) "Trees of Knowledge of Self and Other in Culture: On Models for the Moral Imagination," in L. Rival (ed.) *The Social Life of Trees: Anthropological Perspectives on Tree Symbolism*, Oxford: Berg.

Finkler, K. (2000) *Experiencing the New Genetics: Family and Kinship on the New Medical Frontier*, Philadelphia: University of Pennsylvania Press.

Fleising, U. (2001) "In Search of Genohype: A Content Analysis of Biotechnology Company Documents," *New Genetics and Society*, 20(3): 239–54.

Foucault, M. (1973) *The Birth of the Clinic: An Archaeology of Medical Perception*, Transl. A.M. Sheridan. London: Tavistock.

Franklin, S. and Ragoné, H. (eds) (1998) *Reproducing Reproduction: Kinship, Power, and Technological Innovation*, Philadelphia: University of Pennsylvania Press.

Grétarsdóttir Bender, E.K. (2002) *Íslensk ættfræði* ("Icelandic Genealogy"), BA dissertation, Department of Anthropology, University of Iceland.

Gudmundsson, H., Gudbjartsson, D.F., Kong, A., Gudbjartsson, H., Frigge, M., Gulcher, J.R., and Stefánsson, K. (2000) "Inheritance of human longevity in Iceland," *European Journal of Human Genetics*, 8: 743–9.

Gulcher, J. and Stefánsson, K. (1998) "Population Genomics: Laying the Groundwork for Genetic Disease Modeling and Targeting," *Clinical Chemistry and Laboratory Medicine*, 36: 532–7.

Gulcher, J., Kong, A., and Stefánsson, K. (2001) "The Role of Linkage Studies for Common Diseases," *Current Opinion in Genetics & Development*, 11: 264–7.

Haraway, D. (2000) "Deanimations: Maps and Portraits of Life Itself," in A. Brah and A.E. Coombes (eds) *Hybridity and its Discontents: Politics, Science, Culture*, London: Routledge.

Helgason, A. (2001) *The Ancestry and Genetic History of the Icelanders: An Analysis of MTDNA Sequences, Y Chromosome Haplotypes and Genealogy*, Doctoral dissertation. Institute of Biological Anthropology, University of Oxford.

Helgason, A., Siqurðardóttir, S., Gulcher, J.R., Ward, R., and Stefánsson, K. (2000) "mtDNA and the Origin of the Icelanders: Deciphering Signals of Recent Population History," *American Journal of Human Genetics*, 66: 999–1016.

Ingold, T. (2000) *The Perception of the Environment: Essays in Livelihood, Dwelling and Skill*, London: Routledge.

Ingvarsson, Th., Stefánsson, S.E., Hallgrímsdóttir, I.B., Jónsson jr., H., Gulcher, J., Ragnarsson, J.I., Lohmander, L.S., and Stefánsson, K. (2000) "The Inheritance of Hip Osteoarthritis in Iceland," *Arthritis and Rheumatism*, 43: 2785–92.

Íslendinga sögur og þættir, vols. I–III (1987) Reykjavík: Svart á hvítu.

Kay, L.E. (2000) *Who Wrote the Book of Life? A History of the Genetic Code*, Stanford: Stanford University Press.

Keller, E.F. (2000) *The Century of the Gene*, Cambridge, MA: Harvard University Press.

Klapisch-Zuber, C. (1991) "The Genesis of the Family Tree," *I Tatti Studies: Essays in the Renaissance*, 4(1): 105–9.

—— (2000) *L'Ombre des ancêtres: Essai sur l'imaginaire médiéval de la parenté*, Paris: Fayard.

Laqueur, T.W. (2001) " 'From Generation to Generation': Imagining Connectedness in the Age of Reproductive Technologies," in P.E. Brodwin (ed.) *Biotechnology and Culture: Bodies, Anxieties, Ethics*, Bloomington: Indiana University Press.

Nature (2000) "Manhattan versus Reykjavik," News feature, 406: 340–2.

Newmeyer, F.J. (1986) *The Politics of Linguistics*, Chicago: University of Chicago Press.

Orlove, B.S. (1991) "Mapping Reeds and Reading Maps: Representation in Lake Titicaca," *American Anthropologist*, 18(1): 3–40.

Pálsson, G. (1995) *The Textual Life of Savants: Ethnography, Iceland, and the Linguistic Turn*, Chur: Harwood Academic Publishers.

—— (2002) "The Life of Family Trees and the Book of Icelanders," *Medical Anthropology*, 21 (3/4): 337–67.

Pálsson, G. and Harðardóttir, K.E. (2002) "For Whom the Cell Tolls: Debates about Biomedicine," *Current Anthropology*, 43(2): 271–301.

Pálsson, G. and Rabinow, P. (1999) "Iceland: The Case of a National Human Genome Project," *Anthropology Today*, 15(5): 14–18.

—— (2004) "The Iceland Controversy: Reflections on the Trans-National Market of Civic Virtue," in A. Ong and S.J. Collier (eds) *Global Assemblages: Politics, and Ethics as Anthropological Problems*. Oxford: Blackwell Publishers.

Panofsky, E. (1991) *Perspective as Symbolic Form*, transl. C.S. Wood, Cambridge, MA: Zone Books.

Rabinow, P. (1999) *French DNA: Trouble in Purgatory*, Chicago: University of Chicago Press.

Rheinberger, H.-J. (2000) "Gene concepts," in P. Beurton, R. Falk, and H.-J. Rheinberger (eds) *The Concept of the Gene in Development and Evolution*, Cambridge: Cambridge University Press.

Risch, N. (2000) "Searching for Genetic Determinants in the New Millennium," *Nature*, 405: 847–56.

Rival, L. (1998) "Trees: From Symbols of life and Regeneration to Political Artefacts," in L. Rival (ed.) *The Social Life of Trees: Anthropological Perspectives on Tree Symbolism*, Oxford: Berg.

Schneider, D.M. (1984) *A Critique of the Study of Kinship*, Ann Arbor: University of Michigan Press.

Stefánsson, S.E., Jónsson, H., Ingrarsson, Th., Manolescu, I., Jónsson, H.H., Ōlafsdóttir, G., Pálsdóttir, E., Stefánsdóttir, G., Sveinbjörnsdóttir, G., Frigge, M.L., Kong, A., Gulcher, J.R., and Stefánsson, K. (2003) "Genomewide Scan for Hand Osteoarthritis: A Novel Mutation in Matrilin-3," *American Journal of Human Genetics*, 72: 1448–59.

Strathern, M. (1992) *After Nature: English Kinship in the Late Twentieth Century*, Cambridge: Cambridge University Press.

Wilford, J.N. (2000) *The Mapmakers*, New York: Vintage Books. Revised edition.

Commentaries

10 Maps and mapping practices

A deflationary approach[1]

Sergio Sismondo

Introduction: goals and tools

My main goal in this commentary will be to make a case for a deflationary perspective on maps that emphasizes the multiple interpretations of maps. That stands in opposition to an anti-interpretive stance that sees maps simply as representations of their objects. I would like to start, though, by pointing to the potential contingency of the whole tradition of genetic mapping, a contingency that I think is connected to the main contrast I would like to draw.

The advertising campaigns of New England BioLabs and DoubleTwist.com, mentioned several times during this workshop, capitalize on the excitement of traditional geographic mapping. For the one, an early nineteenth-century map of Africa is projected on a woman in profile, flowing behind her like a large silk veil; where it touches her, Africa turns molecular. For the other, a ship sails along the horizon of tiny Cs, Gs, Ts, and As; the horizon divides another nineteenth-century map of an unnamed shoreline, above, from a space-filling model of DNA, below. In these advertisements the antique maps and the imagery of exoticism evoke the thrill of exploration and discovery, and extend that thrill to genetic exploration and discovery. Clearly, it can extend to genetic mapping at the level of detail of, say, Celera's annotation of its human genome map, or to the Morgan school's early maps. But we might ask, with Angela Creager (Chapter 1, this volume), how genetic mapping at fine levels came to acquire the status of a goal in itself, and an exciting goal to reach. This is one of the respects in which T.H. Morgan and company changed what it was to do genetics; they created a new mapping culture based on the assumption that there was something important to be learned simply from building up complex maps.

In his presentation for this workshop, Stephen Hilgartner (in Chapter 6) describes how public/private boundaries for gene sequences have been constructed. Interestingly, what ends up most obviously on the public side is the genetic map. This is also the case for the "private" effort: Celera ended up, after only a bit of a struggle, putting its map on the web. This suggests that the map itself may not be so interesting after all. We see various hints of that in the other chapters—Alain Kaufmann's (Chapter 7), Gísli Pálsson's (Chapter 9), Soraya de Chaderevian's (Chapter 5)—in which one or another party suggests that one or another map or mapping effort is not worth the effort. Despite its failures, the public/private boundaries might have been easier to negotiate at the edges of maps rather than anywhere else, because genetic maps are not valuable in themselves. Perhaps, as Marcel Weber (Chapter 3) suggested in discussion, in the post-genomic age, genetic maps will become firm tools rather than goals. Reading the material around the Human Genome Project, it appears now that this achievement is more or less behind us, it no longer makes sense as a goal, at least a goal shared by a community.

The mapping cultures of genetics have sometimes led to one another through techniques, suggesting that *Drosophila* was an important point of origin. But in other cases, as Frederic Holmes shows us in his study of Seymour Benzer's fine-structure mapping (Chapter 2), genetic mapping was largely re-invented. More important than continuity of technique, then, may be the idea of mapping, the metaphor that I will argue is bound up with an idea of "brute representation." Genetic mapping seems to have been driven, when it has been driven, by the possibility of maps more than by the easy extension of past mapping practices to current cases. Extension, after all, is not easy. That suggests that without *Drosophila* mapping might not have become an acceptable practice. High-quality maps might never have appeared do-able, and given the effort that they require, they might not have been acceptable as goals for biologists. They might not have become interesting simply as representations of the genome. Genetic mapping, or at least its status, looks a more contingent development than we might think.

What makes maps maps?

Genetic "mapping" has long since ceased to be a live metaphor, and it is not my purpose here to revive its metaphorical status. Nonetheless, it is worth noting how easily the idea of a map can be extended from its core uses to new ones. Almost any spatial depiction of spatial relationships easily can become a map. Spatial depictions of non-spatial relationships become maps with only a little bit more difficulty. And mathematicians can eliminate spatiality altogether in their mappings, making them mere correlations between objects.

What is it that gives "maps" and "mappings" such flexibility? These examples are not in any interesting sense extensions of cartographic practices, which are routinely ignored by metaphorical appropriations of mapping. Cartographic practices can be ignored because maps are thought to be so transparent. Almost any representation can be a map, because maps simply represent.

There is an analogous move in philosophy of science, where the metaphor that scientific theories are maps of the natural world is a recurring one (e.g. Toulmin 1953; Kuhn 1962; Ziman 1978; Azevedo 1997; Giere 1999; Kitcher 2001; Longino 2002). Most uses of map metaphors in philosophy assume that good maps represent their subject matters in a way compatible with a metaphysically modest but globally applicable realism—a "diffuse realism" (Sismondo and Chrisman 2001). The diffuseness of the position allows for a rapprochement between realists and anti-realists. For example, in his *The Philosophy of Science* (1953),

Stephen Toulmin writes,

> Cartographers and surveyors have to choose a base-line, orientation, scale, method of projection, and system of signs, before they can even begin to map an area. ... [T]he alternative to a map of which the method of projection, scale and so on were chosen in this way, is not a truer map—a map undistorted by abstraction: the only alternative is no map at all.
>
> (Toulmin 1953: 127)

The point of this is to steer us away from the idea that there is a non-trivial relation of correspondence between theories and reality, but also to steer us away from the idea that theories might be mere impositions on reality. For Toulmin, this observation should push us in the direction of pragmatism. The initial impetus for that push is, then, a "brute representationalism": maps simply represent.

This is an anti-interpretive stance. However, by turning to practices of cartographic representation, whether genetic or otherwise, we can create new opportunities and imperatives for interpretation. Cartographic practices do not create mere reflections of reality, so there is interesting work to be done to understand abstractions, distortions, instrumentalities, and constructions, and even relations that stand in for correspondence!

For correspondence

Brute representation is intended to obviate metaphysical musings; it is meant to eliminate, for example, correspondence realism about maps. But when we look at how, say, navigational charts are created, we see very precise relations between sightings and maps. For example, Edwin Hutchins (1995) shows how navigation on a US navy ship is a collaborative effort involving sailors, instruments, techniques, and charts. The ensemble of these components does good navigational work because, ideally, each of the components is supposed to produce mechanical responses to the information that is given. While there is no single correspondence relation connecting maps and their objects, there are some very precise and well-understood relations, giving substance to the normally empty notion of a correspondence relation. The criteria for good and bad maps can be extremely stringent. So only in the contexts of cartographic traditions can one say that a depiction is a good or bad map, but within those contexts there *are* good and bad maps.

These stand-ins for correspondence are analogous to the relations that Raphael Falk is concerned about when he talks about the abstract maps of the Morgan school. Falk writes, "Genetic mapping ... started with an abstract notion and provided a map that gave this notion a virtual reality, ... Linkage maps preceded the demonstration of physical, chromosomal reality. It was only later that these virtual maps were proven to have physical counterparts" (Chapter 3 by Falk, volume 1). The mapping procedures initiated by Morgan and his students established a formal relation between rates of recombination and distances on the map. Although this required skill in handling *Drosophila* and seeing mutations, the tedium of sorting, breeding, and counting is something of an indication of the precise relationship between flies and their genetic representation. Correspondence is a similarly precise relationship for more recent genetic maps, and physical maps of DNA. Gene sequencing can now be bought by the base pair, because it has been mechanized. The mechanical relationships between fragments of purified DNA and outputs of strings of Cs, Gs, As, and Ts that have been devised by competing companies are analogous to the procedures that take navigators from sightings to plots on charts.

For instrumentalism

Realist intuitions are aided by the place geometry has in maps and mapping. But geometric fidelity is often abandoned in favor of legibility, ease of use, or some other instrumental value. In philosophical discussions of maps, it is almost obligatory to mention the London Underground map, probably the best-known case of abandonment of geometrical fidelity. The Underground map magnifies distances in the core and shrinks them in the periphery, but not according to any formula. Obviously the map is very successful, as are other subway maps designed along similar lines. The moral is that truth can be sacrificed for utility, and when this sacrifice is made realist interpretations should give way to instrumentalist ones.

The development of technologies to represent topographic features is a different case of instrumentalism. Contour lines represent locations where a particular abstract plane would intersect with the landscape. The proliferation of the contour is not due to some inherent superiority as a visual technique (Wood 1992), but to its instrumentality. It was used by every major military power as a basis for calculations pertaining to artillery, logistics, and military movements; and civil engineers used contours to develop procedures that would estimate the volume of dirt cut-and-fill, drainage, and other properties.

According to Robert Kohler (1994), when Morgan assigned the task of mapping *Drosophila* to his students, it was a classification task. The productivity and mutability of *Drosophila* meant that the neo-Mendelian system of classifying genetic factors was extremely difficult to stabilize: every new mutant changing the description of all previous varieties. Genetic linkage maps were to be a way around that problem, providing a stable classification system for genetic factors. As Kohler points out, being a stable classification system did not amount to being a perfectly realistic representation. Early genetic maps assumed that the rate of crossing-over was constant over the entire length of the chromosome, that the rate of crossing-over directly depended upon distance, and that double crossing-over could be corrected for. None of these assumptions is strictly true. Their correction over the following forty or so years provided part of the contribution of genetic mapping to genetic theory, as Marcel Weber (1998) has neatly argued.

If Kohler is right, then an instrumental interpretation of the Morgan laboratory's maps is a perfectly natural one, even though those maps quickly provided arguments for the reality of genes and their linear ordering (Weber 1998). There is no sharp line between the instrumental and the realistic, so the one can become the other with ease. But some mapping practices and their map products can be usefully interpreted as instrumental, and not simply as attempts at realism.

For constructivism

Cartographic metaphors are increasingly used to speak about the construction of boundaries. Lines set down on maps define territories, or solidify claims to territories. Thus cartography has been a tool of imperialism, helping to establish the extents of empires (Harley 1988). It is also routinely used to establish more mundane boundaries, to locate the sites for new features of the landscape, and to settle controversies (Monmonier 1996). Maps do not, then, simply represent territories, but create them.

Modern genetic mapping is busy creating its own boundaries. As of this, writing patents have been applied for, and mostly granted, on more than 500,000 genes and partial gene sequences. About 35,000 genetic patents are applied for each month. While we should be cautious about saying that these cartographic structures precede territorial ones, the created boundaries are intended to have impacts. Robert Krulwich of ABCNews.com puts it succinctly:

> If you can figure out even roughly what any particular group of these chemicals do in a human being, the United States government will reward you with a U.S. Patent. Once the patent issues, for 17 years you own this strip of DNA, and if scientists want to work on it, you can charge them rent. Even better: if a particular stretch of DNA happens to include part of the cure for cancer or male pattern baldness, for example, then this is very valuable territory. Just the chance to be a DNA landlord has proved to be so desirable that the patent office is now flooded with applications.
>
> (Krulwich 2000)

Constructivism is important to the interpretation of maps in a second way. Among the objects that maps depict most easily are artificial ones. While the edges of a forest may be ill-defined, the edges of a protected forest are usually very carefully specified. The cartography of neatly bounded objects is much simpler than the cartography of fuzzy-edged stuff. Furthermore, it is often a matter of convenience for both cartographers and the users of maps to focus on human structures as markers or signposts. For example, the use of road maps involves looking back and forth between the map and physical markers. For this reason, street names are mildly philosophically interesting, as they function to help make correspondences between maps and territories. Non-natural objects feature prominently on many maps.

Similarly, since the creation of the *Drosophila* experimental system, mutations have played an important role in genetic practice. Mapping mutations is a way of tracking the genetic origins of normal development. Genetics is thus a highly modal science, explaining development in terms of numerous factors not all of which are relevant in any given case—except in the sense developed by genetics. Mutations, of course, are not artificial in the same sense that either a street or a street name is. After all, Theodosius Dobzhansky exploited variation in the wild with some success. But the production of mutations is accelerated in the laboratory, in order to plot the edges of genetically normal development.

For interpretation

I have argued that the relationships between maps and their objects are more complex than the "diffuse realist" would have us believe. David Turnbull's examples (Chapter 8 in volume 1) amplify this point. In so far as mapmakers and map users exploit procedures to link features on maps and features in the world, correspondence realism about maps can make sense. In so far as maps lie in order to facilitate their use, instrumentalism can make sense. In so far as they construct their objects and represent artificial objects, constructivism can make sense. So in place of a single weak interpretation that applies everywhere, I would encourage thinking in terms of what I call "deflationary metaphysics": questions about the nature of representations and their objects arise from within considerations of representational practices, and do not have to be imposed from the outside.

Brute representationalism, then, is a misleadingly simple interpretive (or anti-interpretive) stance. While, as I suggested, it may have helped to establish the place of mapping in the twentieth-century cultures of genetics, it also obscures a number of interesting potential interpretations, ones that the chapters of these volumes have helped to explore.

My quick collection of aspects and uses of maps suggests that we should not try to find a single philosophical perspective, like realism, instrumentalism, or constructivism that makes sense of maps. Such unitary perspectives try to turn interesting and complex objects into dull and simple ones. That is probably wrong for almost any object, but is clearly wrong for maps. At least about conventional maps these points should be relatively unproblematic, because they are entirely mundane. Whether they are so mundane about genetics is less clear.

Acknowledgment

I would like to acknowledge support from Social Sciences and Humanities Research Council of Canada grant 410-00-0515.

Note

1 Central themes in this chapter are discussed in Sismondo and Chrisman (2001).

Bibliography

Azevedo, J. (1997) *Mapping Reality: An Evolutionary Realist Methodology for the Natural and Social Sciences*, Albany, NY: SUNY Press.
Giere, R. (1999) *Science Without Laws*, Chicago, IL: Chicago University Press.
Harley, J.B. (1988) "Maps, Knowledge, and Power," in D. Cosgrove and St. Daniels (eds) *Iconographies of Landscape*, Cambridge: Cambridge University Press, 277–312,
Hutchins, E. (1995) *Cognition in the Wild*, Cambridge, MA: MIT Press.
Kitcher, P. (2001) *Science, Truth, and Democracy*, Oxford: Oxford University Press.
Kohler, R.E. (1994) *Lords of the Fly: Drosophila Genetics and the Experimental Life*, Chicago: University of Chicago Press.

Krulwich, R. (2000) "Genetic Code Landlords: Is Business Getting in the Way of Research?" Available at <http://www.abcnews.go.com/onair/CloserLook/wnt_000228_CL_Genomics_feature.html> (accessed January 28, 2003).

Kuhn, T.S. (1962) *The Structure of Scientific Revolutions*, Chicago, IL: University of Chicago Press.

Longino, H.E. (2002) *The Fate of Knowledge*, Princeton, NJ: Princeton University Press.

Monmonier, M. (1996) *Drawing the Line: Tales of Maps and Cartocontroversy*, New York: Henry Holt & Co.

Sismondo, S. and Chrisman, N. (2001) "Deflationary Metaphysics and the Natures of Maps," *Philosophy of Science* 68 (Proceedings): S38–S49.

Toulmin, S. (1953) *The Philosophy of Science: An Introduction*, London: Hutchinson & Co.

Weber, M. (1998) "Representing Genes: Classical Mapping Techniques and the Growth of Genetical Knowledge," *Studies in History and Philosophy of Biology and the Biomedical Sciences*, 29: 295–315.

Wood, D. (1992) *The Power of Maps*, New York: Guilford.

Ziman, J. (1978) *Reliable Knowledge: An Exploration of the Grounds for Belief in Science*, Cambridge: Cambridge University Press.

11 Mapping

A communicative strategy

David Gugerli

Mapping is an authoritative tool for the production and exploration of relatedness, and it provides its users with a highly efficient analytical instrument for specific problems or questions. Wherever mapping is successfully employed, it allows for the visual aggregation, combination, and interpretation of selected *and* large amounts of data. Hence, mapping is both a compelling visualization technique and a powerful graphical means of orientation within an information domain.

Most communities of scientific practice have been developing or using some sort of maps for some time. That is why Donna Haraway asserts, "Cartography is perhaps the chief tool-metaphor of technoscience" (Haraway 1997: 163). Haraway's phrase has been widely quoted—apparently nobody, however, has dared to ask why cartography or mapping should be called a metaphor. What is so metaphorical about mapping?

Of course, there are many different things that can be mapped, and many different mapping endeavors. There is, for instance, topographic mapping, genetic mapping, cognitive mapping, mapping literature and art, and even the mapping of maps (Harley and Woodward 1987). While some research communities map countries, others map cultures. The only thing that is missing in this abundance of mapping endeavors is the mapping of all forms of mapping.

The proliferation of mapping leaves us, on an analytical level, with two immediate options. We could refrain from using the notion of mapping altogether—where almost everything in scientific practice seems to be related to mapping, the very notion of mapping simply looses its specificity. Alternatively, we might want to introduce an artificial difference between real mapping and metaphorical mapping.

Neither of these is very attractive. While the former is obviously not productive at all, the latter is rather thorny, for the following reason: if we think about topographical mapping as the relevant, the actual, the real, or the original form of mapping, then most of the maps dealt with in this volume are not maps at all. Without noticing, Scott F. Gilbert stumbles over this problem in his provocative statement: "...there *really* are no gene mapping communities in *the actual sense of mapping*. (...) These aren't maps. These are addresses" (Gilbert 2001).[1]

Gilbert's reference to "the actual sense of mapping" is no more helpful than a general assumption of a proliferation of metaphorical cartography in technoscience. One claim is the other's immediate and unavoidable consequence. While the abundance of the metaphorical discourse produces a counter-discourse of the actual, the distinction between real and metaphoric delegitimizes most of the mapping activities that are indeed crucial to many forms of knowledge production.

There is one way out of this vicious circle: the delicate search for functional equivalents in different realms of scientific practice. After all, I still believe that it is useful to conceive of any kind of mapping activity as a part of a generalized form of scientific practice and communication. Of course, we have to be careful not to suggest that problems are identical when they are not comparable. And we have to describe and understand problems in more abstract, fairly general terms, in order to spot the crucial historical similarities and/or differences between various kinds of mapping. Then, and only then, we won't have to talk about mapping in metaphorical terms, or define or construct the "real" or the "actual" form of mapping as a chief-reference point.

The proposed solution to the dilemma has consequences both for the history of genetic mapping and for the history of topographical mapping. The activities involved in mapping landscapes or countries and mapping genes or flies do not fundamentally differ, and they become comparable. The study of one scientific practice illuminates the understanding of the other.[2]

My entry point for the following short and preliminary remarks is Robert Kohler's assertion that "genetic mapping is in principle rather like the triangulation method of topographical mapping." In his noteworthy book *Lords of the Fly*, Kohler writes: "The first step is to establish a baseline by choosing two genes and measuring the distance between them very accurately, by counting large numbers of recombinants. This baseline then serves as a reference to which all other points are related." (Kohler 1994: 65). I would argue that this observation could be radically extended to insights from the history of nineteenth-century topographical mapping.[3] It is my hope that this might also inform specialists in the history of genetic mapping, at least those historians of science who are dealing with Morgan's *Drosophila* group at Columbia University. I have, however, the impression that we will eventually obtain some elements for those mapping practices which have been used throughout the twentieth century in its secular and on-going "hunt for the gene."[4]

Hence, I will expand Kohler's description by shedding light on a few general characteristics of topographical mapping, which can easily be retranslated into the realm of the twentieth-century mapping culture of genetics. Subsequently, I will deal with the following two questions. What are scientists doing when they are mapping? And what effects do maps have on their readers? Answering these questions should enable me to clarify the machinery of a map's visual impact. After all, this evidence is one of the crucial ways that a map accomplishes its communicative tasks.

Goals of the mapping exercise

Which are the most important goals of a nineteenth-century topographical map? In general terms, a topographical map—as a *completed product*—was intended to materialize in a single piece of printed matter the instantaneous visual aggregation of its contents. For special occasions—such as a national exhibition—all parts (or individual sheets) were firmly glued together, put into one frame and presented as one *tableau* to the sponsoring public (Gugerli and Speich 2002).

In order to reach this goal, cartographers maintained that any decent mapping project—in the sense of a surveying *process*—had to start with big entities (the base line and the first-order triangulation), in order to move progressively into ever-smaller units of a country's landscape. In this sequential process of measuring, annotating and (re-)calculating angles or relative distances, on the one hand, and drawing, counting and registering topographical details, on the other hand, an all-encompassing relatedness was produced. Thus, as an *inscription*, the map had to be fine-tuned in every single part of its countless corresponding components. Then, and only then, could maps be judged as "elaborated according to the strong scientific principles that had been observed from the very beginning of a mapping project right up to its accomplishment" (Dufour 1865: 204–9).[5]

With regard to a topographical mapping project's visualizing goals, we can observe how cartographers tried to conventionalize space and landscape. Natural, social, and political entities were systematically subjected to the rules of conventional graphic forms of expression and scanned according to these specifications. At the same time, however, the transformation had to acquire a natural resemblance to the physical world. A map should express, by means of artificial conventions, the highest possible degree of natural similitude. Simple and standardized graphic strokes or minimal pictorial elements should combine into a pictorial effect of a plastically formed landscape seen from an imagined, infinite vertical perspective.

Such an artificial nature or natural artifice had to be achieved in order to guarantee several communicative advantages of the map. The most important advantage was certainly that the apparent scientific, procedural neutrality and abstraction of the map served to mediate interpretative conflicts about a country's or a nation's "nature."

Take as an example nineteenth-century Switzerland: the differences between the dynamically changing urban centers and the increasingly marginalized countryside, the lingering conflicts between Catholic and Protestant regions, the clash between agricultural fears and industrial hopes, the struggle for political participation and representation in an emerging political system called "the nation," not to speak of the huge disparities between languages, cultures, traditions, and social classes. All these and many other tensions were completely ignored by the national survey. The Topographical Atlas did not distinguish the key features of these conflicts; it did not tell its readers about the very distinctive alternatives shaping their future. It rather created a graphically consistent, uniform, and conventionalized landscape, suggesting a homogeneous space of action. When confronted with the Atlas at the 1883 national exhibition, one commentator claimed that everybody involuntarily felt national pride in the defence-worthy glory of the nation as represented by the map. "This object," he continued, "is the pearl of the whole exhibition, and it represents, in a most dignified form, the political unity of Switzerland" (J.v.S. 1883: 269). The map substituted most political differences with topographical subtleties.

Moreover, by means of naturalizing its conventions, a map was also able to hide most of its own constituent predispositions. Thus, a map was not only the astonishingly coherent end-product of a scientific procedure; it was even a quasi-natural picture, which could not be contested on the grounds of individual preferences or political priorities. Wherever possible, "modern" topographical maps successfully refrained from any explicit reference to personal opinions and individual preferences, since they were the product of a collective scientific practice. Due to these moves towards professionalization, many craftsmen in the field of cartography lost their standing in the course of the nineteenth century. Authors like Friedrich Wilhelm Delkeskamp may still have sought to provide their clients with an array of esthetic images of highly individual character for a time; very soon, however, they had to abandon their projects (Delkeskamp 1830–35). The professional community of cooperatively acting topographical engineers successfully absorbed any individual artistry. This aspect of collective authorship strengthened the illusion of cartographic neutrality, and thereby to a large extent augmented a map's legitimation.

Hence, the map of a national survey could stand and mediate the deictic gesture both of an individual and a collective readership. Pointing to the map simultaneously produced indicated presence and absence, difference and identity, specificity and relatedness. Thus, the map reassured individual and collective origins, it showed present positions, and it declared the range of possible future movements.

Cartographic production abounds

What are scientists or engineers doing when they are mapping? Of course, they pursue all—or at least some—of the previously mentioned goals. At the same time, however, they generate preliminary results and side-products as part of their mapping activities. Sometimes they need to change their methods and are forced to adjust the work they have already done. They need to keep records of their measurements and alterations; without an archive for their field notes and previous calculation sheets, they would have to start from scratch when changing the method. In other words, archives help to temporalize the production of a map, to divide it into a series of distinctive processes, which can be—if necessary—regrouped, synthesized and reprocessed (Luhmann 1980). Even if it is true that "gene maps are . . . built up gradually of interconnecting segments from a baseline, just as a topographical map is built up from a baseline in a network of connected triangles" (Kohler 1994: 65), the mapping project has to take into account that the baseline might change its actual (or precisely estimated) length due to a remeasuring of the base or due to the recalculation of the raw-data (Gugerli 1999).

While such adjustments frequently change the expected outcome of a mapping project, preliminary results are also likely to change the assumptions about the object that is being mapped. The genetic landscape of *Drosophila* gradually changed as the Morgan group made progress on its mapping project around the time of the First World War. "Mappers had to create standard *Drosophila* stocks specifically adapted to the peculiar requirements of quantitative measurement. Genetic maps are the blueprints of the standard fly..." (Kohler 1994: 54). Between 1832 and 1865, the same happened to cartographers' assumptions about the form, extension, height, and relative position of the Swiss alpine mountain peaks. Sure enough, Dufour's engineers did not produce standardized mountains, but they labored on the fixation of standard reference points on the top of these mountains. Subsequently, these points could be related to other points for triangulation and projection purposes. Usually, putting something on a map is a long and iterative process, during which this entity gradually gains both specificity and relatedness. Mountains, for instance, become normal, well-defined cartographic entities: they are gradually immersed in a uniform graphical space, which contains numerous other mountains.

Finally, topographic and genetic mapping rearrange their mappers—their position in the project, their influence on the outcome, their organizational relevance, their power of definition, and their body of knowledge. Further, mapping procedures rearrange a whole array of things—nucleotides and rocks, genes and rivers, crossing-overs and triangulation points, theodolites and incubators. While one mapping endeavor gradually stabilized and therefore changed the fly, the other stabilized and reshaped the nation (Gugerli 1998; Gugerli and Speich 2002).

Of course, these similarities could easily be overshadowed by pointing out the differences between the two mapping projects. Nobody would honestly claim that topographical mapping in the era of European nation-building was exactly the same endeavor as genetic mapping during the North-American dawn of the Fordist mass-production era.[6] Nobody would go so far as to ignore the difference between a geneticist's microsope and a topographer's telescope. Quite obviously, no landscape will ever turn into a chromosome. And yet, the problems that have arisen and the strategies that have been developed for genetic and topographic mapping closely resemble each other, at least on an abstract level. Both mapping endeavors produce a similar communication tool by means of observing, measuring, registering, negotiating, recalculating, standardizing, drawing and redrawing, and finally printing, quoting and reprinting. The analytical approach, which looks for functional equivalents, makes genetic mapping at least as real a procedure as topographic mapping. And the same analytical approach shows that cartographers produce as many metaphorical claims for their work as any cultural study on the Morgan group could possibly invent. To put it bluntly: cartographic mapping can also be viewed as a metaphor.

The performance of maps

What are maps doing when they are finished, what is their performance and communicative power? First of all, they serve as filters. They reduce complexity by eliminating differences and evidencing a few selected features. As we have seen, some topographic maps eliminated religious or economic differences between regions. National topographic surveys of the nineteenth century invited their readers to identify unifying aspects—which were to be found in the so-called natural conditions of their country and its landscapes—rather than debate on the spatial expression of highly controversial social differences. The optical homogeneity of a cartographic space, its procedural consistency, and its graphic uniformity stands for the country's political and cultural unity. The genetic distance map of the Morgan group statistically eliminated morphological or phenotypic differences and filtered those genetic elements that were relevant for the genetic description of the ideal type of fly—the standard *Drosophila*.

Second, maps change collective patterns of seeing, collectively shared and standardized ways of perception, and the culturally shaped interpretations, which prevail among their readers. The members of the Fly Room at Columbia University gradually got acquainted with the graphic representation of *Drosophila*'s genetic conditions. They learned how to read and see the genetics of "The Fly"—as well as how to manipulate it—by studying such an abstract graphical tool as a chromosome map. Similarly, readers of a national topographical survey in the nineteenth century learned how to read, how to see, and how to manipulate or act within the natural, the institutional, and the ideological framework of the nation.

Third, maps always serve as tranquilizers against the *horror vacui*, the fear of the void, by suggesting that they provide an optically consistent space, a graphically uniform representation. This is what I would call the fiction of completeness of cartography. Wherever the (unfinished) map contains a white spot or an obvious gap between two addresses, the mapping practitioners have to fill it with some information. Maps ask to be completed. It is noteworthy that the Human Genome Project successfully turned this imperative into a strong argument for obtaining the necessary financial resources and the required institutional support: "Starting maps and sequences is relatively simple; finishing them will require new strategies or a combination of existing methods. After a sequence is determined...the task remains to fill in the many large gaps left by current mapping methods."[7] No matter what purpose completing the human genome map should serve—a decent map is simply not allowed to have unknown regions.

Following this, mapping projects integrate different levels of precision into one representational space. They distribute, as it were, accuracy and precision over their entire graphematic space. Their highly celebrated sequences of data transpositions, which eliminate individual authorship and create newly imagined communities or identities, unify both their objects and their readers. This is probably the most important aspect of a politics of truth provided and performed by any kind of scientific mapping.

Generating social and technical evidence

Finally, we have to consider the sociotechnical evidence a map is able to generate, since it is the collective evidentiary value of a map that enhances its communicative functions. Much more than procedural consistency alone has to be accomplished in order to cartographically produce the visualization of a nation's collective space of action, of cultural reproduction, of political planning, public administration, or civil engineering. Curiously enough, the conditions of cartographic evidence can be found both in a map's making (through the production and expression of relatedness) and in its final appearance (through the power of visual aggregation).

There is a whole array of conditions that produce and eventually stabilize a map's evidentiary value in its final expression. Since maps have to be produced following the most exact scientific methods, it is necessary to make public the standards, procedures, and the level of precision of a mapping project. The exponents of nineteenth-century national surveys became eager to publish some of their data in an internationally standardized form, since this allowed for their scientific validation. This was especially true for data concerning base measurements and first-order triangulations as all the other data were much too clumsy and could not be presented to anybody outside the cartographers' study. Publishing preliminary results, which provided evidence of the scientifically consistent progress of the genetic mapping of *Drosophila* or the human genome, was probably as important as the publication of any kind of final map or final text.[8]

Nevertheless, mapping projects always have to acquire a huge amount of cultural sanctioning through the presentation of their final product at national and international exhibitions, or through the publication of popular accounts and biographies, and it is absolutely crucial for a map to achieve a cartographical conditioning of its readers. In other words, scientific and cultural approval co-produced a map's visual power.

Part of such approval is the seminal accounts of the very conditions of the mapping practice. These remain decisive for the production of a map's evidentiary value even after the map is finished. The forms of collaboration between political entities as well as the division of labor among the many participants of a project have to be carefully organized *and* discursively connected with a mapping project's product. Thus, it becomes important for the value of a map as a piece of evidence to declare the participation of molecular biologists or astronomers, venture capitalists, universities, governments, or national scientific foundations, software engineers or engravers, laboratory assistants or topographers. Seemingly, the linking of the spatial visualization with further scientific, administrative, and technological practice usually enhances a map's value as a means of communication and a piece of evidence.

Focusing on different mapping projects in terms of functional equivalents sheds light on the fact that the negotiation of representational means and formats, the fine-tuning of a map's conditions of production, the history of its development, and even the self-representation of the mapping endeavor are of great importance when it comes to understanding a map's persuasiveness. Without agreement and final decision on these issues, a map will never acquire a collectively sanctioned visual power value and it will never be able to serve its communicative functions, either among its producers or its readers. This self-reinforcing tendency of a mapping project might even go so far as to put into oblivion some of the practical communicative purposes that initiated the project. The production of relatedness and visual aggregation that serves the ends of governments and the goals of scientific communities develops a dynamic, which goes far beyond any explicit instrumental utility. Thus, a map mediates communicative processes, which were never planned, and sometimes not even imagined, by its own, particular mapping community.

Notes

1 My emphasis.
2 "Il n'y a pas de phénomènes fondamentaux. Il n'y a que des relations réciproques et des décalages perpétuels entre elles" (Foucault 1994: 277).
3 I am drawing upon research carried out over the last few years together with Daniel Speich. I thank him for invaluable discussion time. Claims, errors, omissions, and shortcomings are my responsibility. See Gugerli and Speich (1999, 2002).
4 See "Human Genome Project Information, Mapping and Sequencing the Human Genome" on <http://www.ornl.gov/hgmis/publicat/primer/prim2.html> for its astonishing, late-twentieth-century parallels to what the following remarks try to describe with reference to two older mapping endeavors. For an overview on the hunt for the gene see Keller (2001); Kay (1999); Nelkin and Lindee (1995).
5 My translation.
6 "The standard maps of 1919–1923 represented data from some ten million flies, and altogether about thirteen to twenty million flies were etherized, examined, sorted, and processed!" (Kohler 1994: 67).
7 Human Genome Project Information, Mapping and Sequencing the Human Genome, <http://www.ornl.gov/hgmis/publicat/primer/prim2.html>.
8 See "Human Genome Project Information" <http://www.ornl.gov/hgmis/project/progress.html>. Several other websites continuously monitored the Human Genome Project's sequencing output.

Bibliography

Delkeskamp, F.W. (1830–35) *Malerisches Relief des klassischen Bodens der Schweiz. Die Delkeskamp-Karte; topographisch-künstlerisches Bild der Urschweiz aus den Jahren 1830–1835. Nach der Natur gezeichnet und radiert von Friedrich Wilhelm Delkeskamp*; kommentiert von E. Imhof (1978), Dietikon, Zürich: Stocker.

Dufour, G.-H. (1865) "Schlussbericht des Herrn General Dufour über die topographische Karte der Schweiz vom 31. Dezember 1864," *Bundesblatt der schweizerischen Eidgenossenschaft*, 17: 203–14.

Foucault, M. (1994) *Dits et écrits 1954–1988 par Michel Foucault, Tome IV. 1980–1988*, in D. Defert and F. Ewald (eds), Paris: Gallimard, 270–85.

Gilbert, S.F. (2001) "Fate Maps, Gene Expression Maps, and the Evidentiary Structure of Evolutionary Developmental Biology" paper presented at the conference *The Mapping Cultures of Twentieth Century Genetics*, Berlin, Germany, March 1–4, 2001.

Gugerli, D. (1998) "Politics on the Topographer's Table: The Helvetic Triangulation of Cartography, Politics, and Representation," in T. Lenoir (ed.) *Inscribing Science: Scientific Texts and the Materiality of Communication*, Stanford: Stanford University Press, 91–118.

—— (1999) "Präzisionsmessungen am geodätischen Fundament der Nation. Zum historischen Anforderungsreichtum einer vermessenen Landschaft," in D. Gugerli (ed.) *Vermessene Landschaften. Kulturgeschichte und technische Praxis im 19. und 20. Jahrhundert*, Zürich: Chronos, 11–36.

Gugerli, D. and Speich, D. (1999) "Der Hirtenknabe, der General und die Karte. Nationale Repräsentationsräume in der Schweiz des 19. Jahrhunderts," *WerkstattGeschichte*, 23: 61–82.

Gugerli, D. and Speich, D. (2002) *Topografien der Nation. Politik, kartografische Ordnung und Landschaft im 19. Jahrhundert*, Zürich: Chronos.

Haraway, D.J. (1997) *Modest_Witness@Second_Millenium. FemaleMan©_Meets_Oncomouse™*, New York: Routledge.

Harley, J.B. and Woodward, D. (eds) (1987) *The History of Cartography*, Chicago, IL: University of Chicago Press.

Human Genome Project Information. Available at <http://www.ornl.gov/hgmis/project/progress.html> (accessed February 27, 2002).

Human Genome Project Information, Mapping and Sequencing the Human Genome. Available at <http://www.ornl.gov/hgmis/publicat/primer/prim2.html> (accessed February 27, 2002).

J.v.S. (1883) "Die Landesausstellung in militärischer Beziehung," *Allgemeine Schweizerische Militär-Zeitung. Organ der schweizerischen Armee*, 29: 269–329.

Kay, L.E. (1999) "In the Beginning Was the Word?" in M. Biagioli (ed.) *The Science Studies Reader*, London, New York: Routledge, 224–33.

Keller, E. Fox (2001) *Das Jahrhundert des Gens*, Frankfurt a. M., New York: Campus.

Kohler, R.E. (1994) *Lords of the Fly. Drosophila Genetics and the Experimental Life*, Chicago, IL: University of Chicago Press.

Luhmann, N. (1980) "Temporalisierung von Komplexität. Zur Semantik neuzeitlicher Zeitbegriffe" in N. Luhmann (ed.) *Gesellschaftsstruktur und Semantik*, vol. 1 Frankfurt a. M.: Suhrkamp, 235–300.

Nelkin, D. and Lindee, M.S. (1995) *The DNA Mystique. The Gene As a Cultural Icon*, New York: Freeman.

Index

For Product Safety Concerns and Information please contact our EU
representative GPSR@taylorandfrancis.com Taylor & Francis Verlag GmbH,
Kaufingerstraße 24, 80331 München, Germany

Printed and bound by CPI Group (UK) Ltd, Croydon, CR0 4YY

01/05/2025

01858548-0001